Advances in Rotary Kiln Sponge Iron Plant

Advances in Rotary Kiln Sponge Iron Plant

K.K. Prasad
B.Sc., Hons. (Chem.), B.E., Ph.D. (Mech. Engg.)
Formerly, Deputy General Manager
Research and Development Centre for Iron and Steel
Steel Authority of India Limited, Doranda, Ranchi

Hem Shankar Ray
B.Tech., Hons. (Metallurgical Engg.), M.A.Sc., Ph.D.
Emeritus Scientist
Central Glass and Ceramic Research Institute
Kolkata

PUBLISHING FOR ONE WORLD

NEW AGE INTERNATIONAL (P) LIMITED, PUBLISHERS
New Delhi • Bangalore • Chennai • Cochin • Guwahati • Hyderabad
Jalandhar • Kolkata • Lucknow • Mumbai • Ranchi
Visit us at www.newagepublishers.com

Published by New Age International (P) Ltd., Publishers
First Edition: 2009

Branches:

- 36, Malikarjuna Temple Street, Opp. ICWA, Basavanagudi, **Bangalore**. ✆ (080) 26677815
- 26, Damodaran Street, T. Nagar, **Chennai**. ✆ (044) 24353401
- Hemsen Complex, Mohd. Shah Road, Paltan Bazar, Near Starline Hotel, **Guwahati**. ✆ (0361) 2543669
- No. 105, 1st Floor, Madhiray Kaveri Tower, 3-2-19, Azam Jahi Road, Nimboliadda, **Hyderabad**. ✆ (040) 24652456
- RDB Chambers (Formerly Lotus Cinema) 106A,1st Floor, S.N. Banerjee Road, **Kolkata**. ✆ (033) 22275247
- 16-A, Jopling Road, **Lucknow**. ✆ (0522) 2209578
- 142C, Victor House, Ground Floor, N.M. Joshi Marg, Lower Parel, **Mumbai**. ✆ (022) 24927869
- 22, Golden House, Daryaganj, **New Delhi**. ✆ (011) 23262370, 23262368

ISBN : 978-81-224-2401-0

Rs. 495.00

C-09-02-3441

Printed in India at Glorious Printers, Delhi.
Typeset at Inhouse.

PUBLISHING FOR ONE WORLD
NEW AGE INTERNATIONAL (P) LIMITED, PUBLISHERS
4835/24, Ansari Road, Daryaganj, New Delhi-110002
Visit us at **www.newagepublishers.com**

Preface to the First Edition

A number of books on sponge iron are available and they deal largely with gas based sponge iron making. There is no comprehensive treatise available on coal based rotary kiln sponge iron making. This is in keeping with the global popularity of gas based sponge iron making which accounts for over 90% of the total sponge iron production. But the situation in India is different. Locational conditions here are such that, except for a limited area in Western coast where regular availability of natural gas at competitive price can be ensured, the enhancement of sponge iron making capacity has been through the coal based rotary kiln route.

But all is not rosy for the rotary kiln. Even though rotary kiln is a versatile reactor and is able to handle raw materials of widely varying properties, but this statement presupposes the prior knowledge of variation. Apart from this high sensitivity to variation in raw material characteristics, there are the problems of high-energy consumption and low productivity. Before long the technology must improve itself to make it more energy efficient and environment friendly.

The present book focusses on the optimised process concept, which has been developed, based on the extensive R&D work undertaken in the Direct Reduction Process Development (DRPD) Division, which is under the Research and Development Centre for Iron and Steel (RDCIS) equipped with a fully functional Rotary Kiln Sponge Iron Pilot Plant and associated laboratories with facilities for Process Control.

Thus we have highlighted the recent advances in various features of rotary kiln. To make the text comprehensive a chapter on reaction kinetics has been included as also a section on iron ore and pellets. Other supporting sections are on aerodynamics, project engineering, auto-ignition, coal throwing etc.

We hope readers would appreciate this endeavour of ours and help us by offering constructive criticism.

K.K. Prasad
H.S. Ray

Acknowledgement

The authors would like to thank foremost RDCIS SAIL; the first author compiled a very significant amount of information presented in this book during his tenure here between 1981 and 2006. Others include employees of DRPD Division of RDCIS SAIL, other employees who assisted the project, both within SAIL and from other organisations like HEC and professors and scientists in IIT's (Mumbai, Kanpur, Kharagpur, IT-BHU) and RRL's.

Our special thanks go to Dr. Asim Kumar Ray of RDCIS, SAIL, to provide help and support in bringing out this book. We would also like to thank Prof. R.C. Gupta of IT, BHU for making available macro- and micro-graphs of sponge iron as also a few samples and some requisite information.

A number of rotary kiln sponge iron plants of India have provided a good deal of information regarding their operation, some of which we have incorporated in this write up. We are thankful to the management of all such plants.

GLOSSARY

(Important Terms, Symbols, Abbreviations and Notes)

IMPORTANT TERMS

Decrepitation	=	Tendency to break into smaller sizes (degrade) generally on heating and/or during reduction.
Free-board	=	Open space in the reactor above the solid charge.
Granular material	=	Solid material which is not very fine (i.e., not ground) and is in a relatively close size range (e.g., 10 to 40 mm, 6 to 20 mm, 5 to 18 mm, etc.).
Mole, Molar	=	Molecular weight expressed in weight measures (e.g., grams – g-moles; kilograms – kg-moles).
Reducibility (of iron ore)	=	The ease with which the ore or oxide gets reduced in presence of reducing gases – Described in Chapter 9.
Reactivity (of coal char)	=	The ease with which the coal or char regenerates carbon monoxide by reaction with either CO_2 (Carboxy–reactivity), O_2 (Oxy-reactivity) or H_2O (Hydroxy-reactivity) – Described in Chapter 9.
Sponge Iron, Direct Reduced Iron (DRI)	=	Iron produced by the reduction of iron oxide (ore) in solid state. Since the process is accompanied by large reduction in volume (53%), a lot of pores appear in the metallic product, which looks like sponge under the microscope.

SYMBOLS

Δ	=	Denotes changes in properties (e.g. ΔG)
ϕ	=	Diameter
α	=	Angle of Repose
^{o}	=	Denotes Standard State (e.g. ΔG^{o}); also denotes degree (temperature, e.g., °C, °F, °K, °R)
ϕ	=	Filling Angle (Semi vertical angle subtended by charge bed on the kiln axis – Chapter 6)
ψ	=	Angle of Inclination (of kiln)
γ	=	Angle between surface of material and kiln axis (Zero for uniformly loaded kiln)
G	=	Gibbs Free Energy
H	=	Enthalpy
P	=	Partial Pressure

S	=	Entropy
T	=	Temperature (Absolute – Degrees Kelvin or Rankine)
t	=	Residence Time

ABBREVIATIONS

ABC	=	After Burning (and Cooling) Chamber
ACAR	=	Allis Chalmers Agglomerated Reduction
ACCAR	=	Allis Chalmers Controlled Atmosphere Reduction
BASL	=	Bihar Alloy Steels Limited
BF	=	Blast Furnace
BIS	=	Bureau of Indian Standard
BOD	=	Biological Oxygen Demand
BOF	=	Basic Oxygen Furnace
C	=	Centigrade, Celsius (temperature)
Cal	=	Calories
C_{fix}	=	Fixed Carbon (also FC)
C_p	=	Heat Capacity at Constant Pressure (Specific Heat of one mole)
CFD	=	Computational Fluid Dynamics
Co.	=	Company
COD	=	Chemical Oxygen Demand
CODIR	=	Coal Ore Direct Iron Reduction (A sponge iron making process)
CONARC	=	Converter & Electric Arc composite furnace
CPT	=	Crossing Point Temperature
Daf	=	Dry and ash-free
Dia	=	Diameter
DIN	=	Deutsches Institute fur Normung e.v. (Standardisation institute for Germany)
Dmf	=	Dry and mineral matter free
DNA	=	Deoxyribo-Nucleic Acid
DPR	=	Detailed Project Report
DR	=	Direct Reduction
DRC	=	M/s Direct Reduction Corporation
DRI	=	Direct Reduced Iron
DRPD	=	Direct Reduction Process Development Division (of RDCIS, SAIL)
EAF	=	Electric Arc Furnace
F	=	Fahrenheit (temperature)
FC	=	Fixed Carbon

Fe(M), Fe(m)	=	Metallic Iron
Fe(T), Fe(t)	=	Total Iron
Finmet	=	Fine Ore Metallisation (A sponge iron making process developed from (FIOR)
FIOR, Fior	=	Fine Iron Ore Reduction (A sponge iron making process using fluidised bed)
FR	=	Feasibility Report
FY	=	Financial Year
GCal	=	Giga Calories (10^9 Calories)
HBI	=	Hot Briquetted Iron
HEC	=	M/s Heavy Engineering Corporation
HEG	=	M/s Hindustan Electro Graphite
HSL	=	M/s Hindustan Steel Limited
HyL	=	Hojalata ya Lamina, or Sheets and Plates (also a sponge iron making process developed by M/s HyL, Mexico)
IDT	=	Initial Deformation Temperature
IF	=	Induction Furnace
IISCO	=	M/s Indian Iron and Steel Company Limited
IIT	=	Indian Institute of Technology
IS	=	Indian Standards (formulated by BIS)
IT, BHU	=	Institute of Technology, Banaras Hindu University (Varanasi, India)
JIS	=	Japanese Industrial Standard
JSPL	=	M/s Jindal Steel and Power Limited
JSW	=	M/s Jindal South West Limited
JVSL	=	M/s Jindal Vijayanagar Steel Limited (now JSW)
K	=	Kelvin (absolute temperature)
KCal, kCal	=	Kilocalories
KVA	=	Kilo-Volt Ampere
LOI	=	Loss on Ignition
m	=	metre, metres
mm	=	millimetre, millimetres
Midrex	=	A sponge iron making process developed by the then Midland Ross Company
MVA	=	Mega Volt Ampere
NALCO	=	National Aluminium Company Limited
NMDC	=	National Mineral Development Corporation
NML	=	National Metallurgical Laboratory

NOx	=	Oxides of nitrogen
OSIL	=	M/s Orissa Sponge Iron Limited
QRT	=	Quick Response Thermocouple
R	=	Rankine (absolute temperature)
RDCIS	=	Research and Development Centre for Iron and Steel (of SAIL)
RK	=	Rotary Kiln
rpm, RPM	=	Revolutions per minute
RRL	=	Regional Research Laboratory
R&D	=	Research and Development
SAF	=	Submerged Arc Furnace
SAIL	=	M/s Steel Authority of India Limited
SIIL	=	M/s Sponge Iron India Limited
SL/RN	=	Stelco, Lurgi, Republic Steel and National Lead (A sponge iron making process)
SOx	=	Oxides of Sulphur
SPM	=	Suspended Particulate Matter
TDR	=	TISCO (or Tata) Direct Reduction (A sponge iron making process)
TISCO	=	M/s Tata Iron and Steel Company Limited (Tata Steel)
tpa/tpy	=	Tonnes per annum or year
tpd	=	Tonnes per day
UHP	=	Ultra High Power
VM	=	Volatile Matter

NOTES

1. For arithmetic operation *(Star) has been used to denote multiplication to avoid confusion with alphabet x. Also / (slash) has been used to denote division to avoid increasing line space.
2. Italics in text generally denote the reference in the Bibliography section.
3. CGS units have been used in the text and not SI units, as is the current trend. One important conversion factor, which is required for linking SI and CGS systems is that, one calorie equals 4.186 joules.

List of Figures

List of Tables

Contents

1

CHAPTER

Introduction

1.1 WHAT IS SPONGE IRON AND WHAT IS DRI?

Sponge iron, also called Direct Reduced Iron (DRI), is formed when an oxide of iron is reduced to its metallic form below the melting points of the metal and the oxide. Naturally, the shape of oxide is retained in the reduced form. Since there is no melting, block volume remains practically unchanged. As weight is reduced and true density increases, true volume is reduced considerably. These opposing factors are reconciled by the formation of a lot of microscopic pores, and under the microscope it looks like sponge and

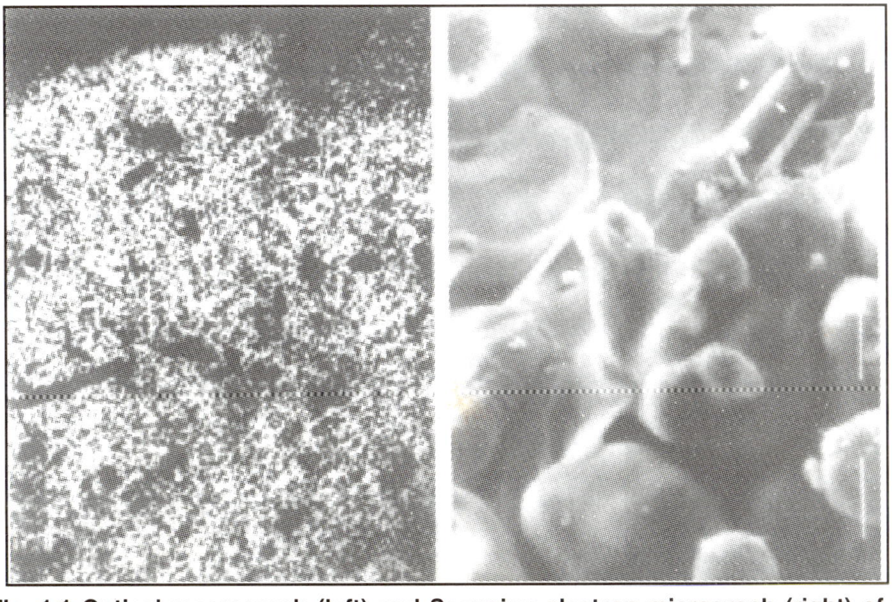

Fig. 1.1 Optical macrograph (left) and Scanning electron micrograph (right) of a typical sponge iron sample

(Courtsey Prof. R.C. Gupta, IT-BHU)

hence its name (Fig. 1.1). Unlike the name, it is not at all soft to the touch, but a well-prepared sponge iron would be much softer under the hammer than normal iron or steel.

The terms 'Direct Reduced Iron' and 'Direct Reduction (DR)' have a historical background. In the present context, DR signifies reduction without melting, which may appear odd to us now, but it is not so if we look through the eyes of the earlier (primitive to us) iron and steel makers.

Sponge iron was the first extracted iron which man had produced (Fig. 1.2). The first extraction was probably an accident, but the art was later refined to make high-grade iron, almost pure, followed by solid-state carburisation (cementation – this is also the principle behind the case hardening process of the presen+t day). The steel produced were of finest grades, the artisans were able to control the chemistry (without knowing the chemistry) and heat treatment with precision.

A story is being told by the indigenous people of the Bastar region in central India according to which iron was smelted first

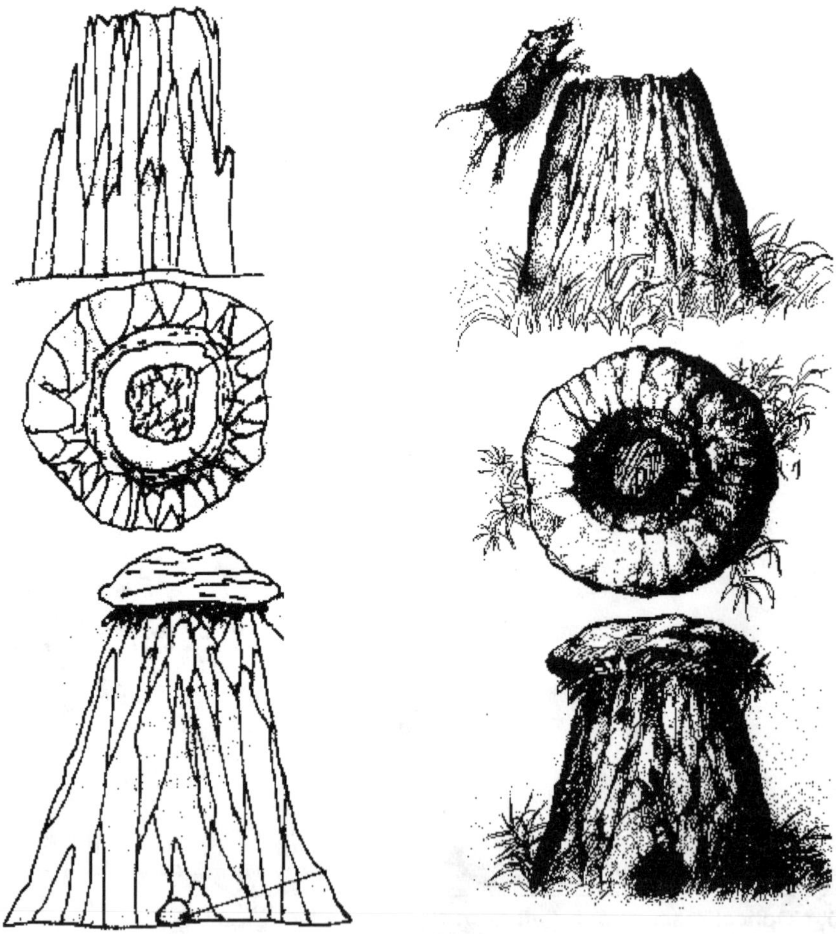

Fig. 1.2 Accidental smelting of iron in an anthill

accidentally in an anthill. To catch a rat which has taken refuse inside the hollow interior of an anthill with a damaged top (it also housed a tree stub) a hunter covered the top with a red rock (!) using dry leaves around to seal all openings at the top. He then made an opening at the base, in a direction where wind was favourable, to blow in the fire he made. The suffocating rat tried to come out and was caught. But overnight the favourable wind intensified the fire and next day an object was found which was much better for their

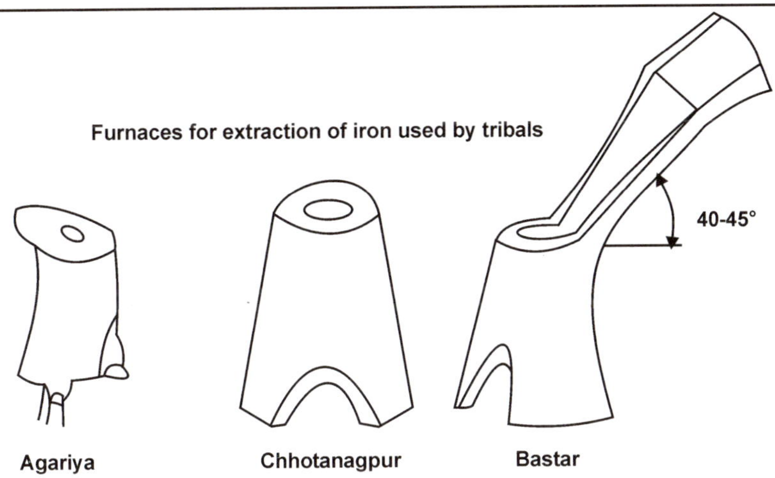

Fig. 1.3 Some primitive iron making furnaces of India

Fig. 1.4 Primitive iron making practice of Orissa

Fig. 1.5

hunting weapons than anything they had used before. (Fig. 1.2–On the left is the original sketch of the authors. The right half is as it has been reproduced in the book 'Ferrum Hunters' by Das & Ray–2005).

When man could find means of attaining higher temperatures, he could make much more iron from furnaces of same capacity through the liquid iron route. In this process much more carbon was picked up, which had to be removed during further processing to make steel, and this obviously was an indirect process route compared to the traditional method of making a pure form of iron and carburising it to make steel. By comparison the traditional method come to be known as the Direct Reduction (DR) process and the iron produced by reduction in solid state as DRI (Sponge Iron).

Sponge iron was thus originally produced using coal or charcoal in a mud shaft

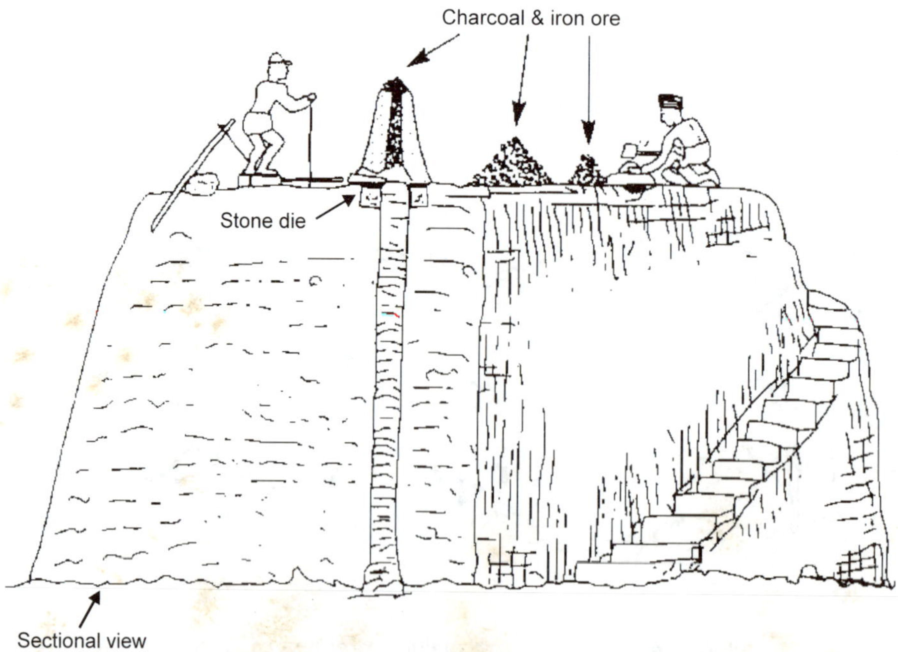

Fig. 1.6 A view of the enigmatic Iron Pillar and a probable method of its making

(Figs. 1.3 and 1.4) and this turned out to be the forerunner of the modern day blast furnace for producing liquid iron.

India has many iron based ancient monuments with remarkable properties. The most famous of these is the Iron Pillar of Delhi (Fig. 1.5). Iron for all these monuments and artefacts were extracted through the sponge iron route – made in mud- or clay-shaft furnace. Some of the indigenous people in Chhattisgarh and Jharkhand States are still practising this technique. Studies of the products made by these people, as well as study of ancient heritage (Delhi Pillar being one), have revealed some points of importance.

1. It was easy to make very low carbon wrought iron by such a process. While higher carbon iron could be attempted, control of carbon was found difficult. Large variation of carbon between subsequent heats was observed.

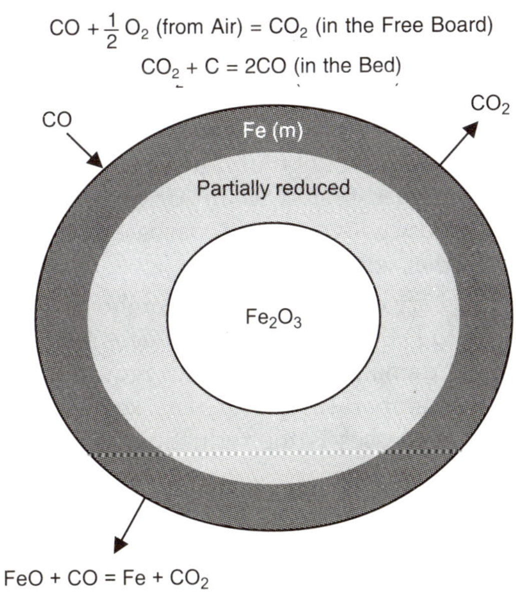

Fig. 1.7 Layer-wise reduction of iron oxides in a rotary kiln

2. All the samples examined by us as well as those reported in literature contained either no manganese or in traces only.
3. As there was no control over phosphorous in the process, high phosphorous in the iron ore led to high phosphorous wrought iron. Probably such an ore was deliberately chosen for the Delhi Pillar.
4. The Iron Pillar of Delhi was probably made by making high carbon 'blooms' successively one over the other, and beating the freshly formed 'bloom' on to the older layer–while keeping a split hollow cylindrical stone die around (Fig. 1.6). This way the bloom layers could be effectively welded to each outer, but left the surface extremely rough. The erect iron mass was then taken down, heated and rotated on a manually driven stone lathe and forged between stone dies–much like the modern forging of railroad axles. During this forging process the surface got heavily decarburised while the interior remained relatively high in carbon.
5. Studies conducted by Prof. Balasubramaniam of IIT Kanpur have indicated that a combination of extremely low manganese and high phosphorous has contributed to the excellent corrosion resistance of the Iron Pillar. Too high a carbon level would have had an adverse effect on this property.

But, now a days, it is found commercially feasible to produce sponge iron in a shaft by using reducing gases produced in a separate reactor. On the other hand, sponge iron production with direct use of coal is effected predominantly in rotary reactor or kiln, which is our present focus. The reduction process in a rotary kiln is also used, or is feasible, for a few other oxides or ores which have been touched upon later.

1.2 HISTORY OF MODERN SPONGE IRON MAKING

Sponge iron making played a dominant role in the production of metallic iron before 1800 A.D., but could not compete with blast furnace technology in terms of efficiency and economy. However, in recent times, constraints relating to raw materials and intermediate inputs to steel making have re-stimulated interest in the development of technically sound and economically viable high tonnage sponge iron processes. As a result, various processes have been developed, tested and adopted throughout the world and presently sponge iron accounts for over 8% of the total iron production in the world. Although the share of coal-based variant is less than 2%, the upward trend, especially in India, is very promising.

Iron metal and its oxides have melting points close to each other and all are about 1500°C or above. However, there is a tendency for formation of 'clusters,' 'agglomerates' and 'accretion' or 'ring formation' during actual manufacture of sponge iron in the temperature range of 900 to 1100°C. Ring formation is a phenomenon occurring exclusively in a rotary kiln, while clusters and agglomerates are common to both rotary kiln and shaft processes. In the present context we cannot conceive of a reactor to manufacture sponge iron, which operates above 1100°C, even though there have been prolonged attempts to produce sponge iron and semi-fused iron at higher temperatures. The Krupp Renn process of production of iron "luppen" in Germany between First and Second World Wars is an example of one such attempt which had met with partial success. Heavy deposit formation could not be prevented and the process had to be abandoned on this account.

When rotary kiln processes could not come up to the expectation to meet the iron needs of the industry, attempts were made to reduce iron oxide in shafts using gases generated externally. In the Wieberg Soderfors process, attempt was made to generate reducing gas through the gasification of coke using external energy. But the first truly viable shaft process of sponge iron making was developed by M/s Hojalata ya Lamina (meaning Sheets and Plates) which reduced iron ore pellets in stationary bed using externally reformed natural gas (HyL process). Later on they converted the set of stationary beds to a moving bed reactor, but Midland Ross Company did the same earlier and got a head start in sponge iron making market and gained a lead, which they are maintaining even today (Midrex process). But rotary kiln was the only choice in the regions, where natural gas availability was restricted. Consequently, sponge iron making in rotary kiln flourished first in South Africa, and now India produces over 70% of coal based sponge iron in the world.

It is very easy to reduce the higher oxides of iron to FeO stage. One only needs to meet the heat demand, the reductant requirement and the temperature level. The key step in all DR processes is the reduction of FeO to metallic iron stage.

1.3 IRON ORE REDUCTION AT A GLANCE

If we expose an iron oxide pellet or lump to reducing gases such as CO and H_2 at an adequate temperature, sponge iron is the product. If we cover iron oxide with coal or charcoal and heat in a muffle furnace, the result is same. One may be inclined to think that physical contact with carbon should cause reduction, but the interior of iron oxide lump or pellet also gets reduced, which is not in contact with carbon. Further, if we try to carry out the same exercise under progressively reduced pressures, the rate of reduction progressively reduces. It is obvious, therefore, that reduction is effected by reducing gases, even if we keep iron oxide and coal in contact. For facilitating quicker regeneration of these reducing gases from coal, an inclined rotary furnace was found handy and convenient.

Figure 1.7 schematically represents how iron ore pellet – or for that matter a lump – is gradually reduced in layers when exposed to rotary kiln conditions. A pellet of satisfactory metallisation would not have any core of Fe_2O_3 left in it, while the partially reduced layer would be restricted to a very limited area in the core.

In Fig. 1.8 we have tried to present pictorially the mechanism of reduction and pore formation during sponge iron making.

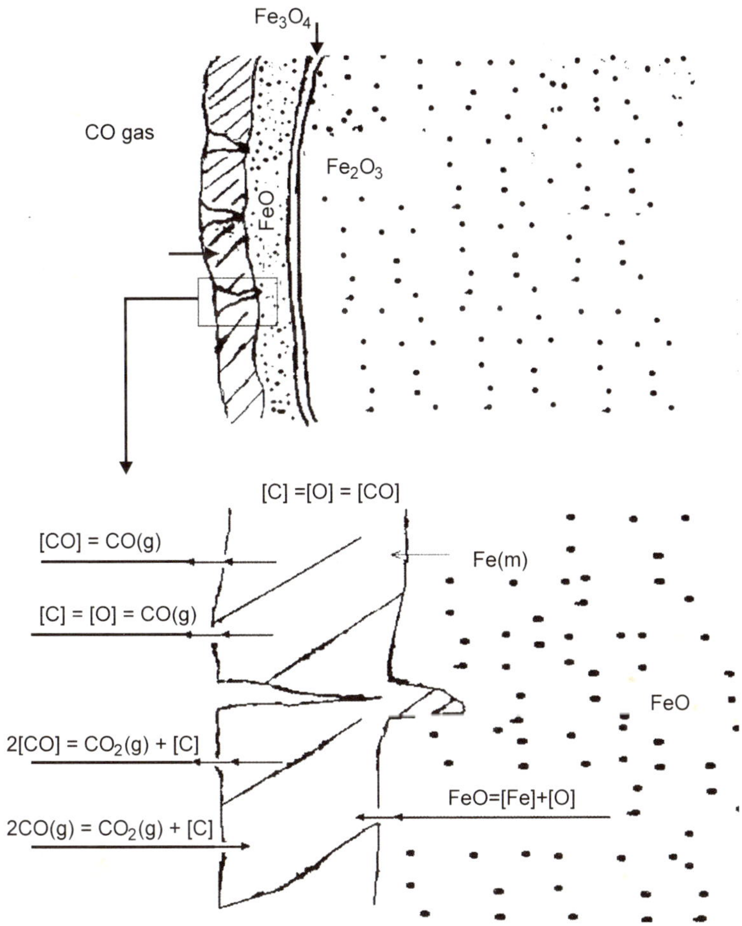

Fig. 1.8 A possible mechanism of reduction and pore formation during sponge iron making in rotary kiln

1.4 MULTIPLE ROLE OF A ROTARY KILN

The use of rotary kiln is limited not only to produce DRI but it is most commonly used for calcination of lime and refractory material, production of cement and also for pre-treatment (pyro-metallurgical beneficiation) of some non-ferrous ores and waste oxides.

These kilns are comparatively longer and of smaller diameter (*i.e.*, of higher L/D ratio) since heat transfer from flame to the solid charge is the main requirement. Because of its versatile usage in processing materials for drying, heating, calcining, agglomerating, sintering, reducing, melting, etc., the rotary kiln will continue to play an important role in different industries in future too.

Rotary Kiln Process of Making Sponge Iron

2 CHAPTER

2.1 HISTORICAL BACKGROUND

The production of steel began in ancient times; but because of the complexity and slow speed of the ancient process, they could not be carried out on a very large scale. Consequently, they were replaced by the high production rate 'indirect process,' and the development of modern DR Process did not begin until the middle of 19th century.

Perhaps the very first patent in U.K. for sponge iron making was in 1792 presumably using a rotary kiln. More than 100 DR processes have been invented and operated since 1920. Most of these have died down. But some of them have re-emerged in slightly different form.

As touched upon earlier, sponge iron is mainly produced from ore by two different routes – (a) by reducing gases (CO and H_2) in a shaft furnace, and (b) through direct treatment with coal in a rotary kiln.

2.2 IMPORTANT FEATURES

The coal based rotary kiln process of making sponge iron is the focus of the present write up. Although many different processes and process concepts have been emerging in this area, there were rapid births and deaths of these processes and process concepts in the middle of the twentieth century. But those operating successfully at present have many features in common. Some of the common or slightly differing features are:

(i) System of sealing to prevent air ingress into the reactor,

(ii) System of throwing or slinging coal from discharge end of reactor,

(iii) System of weigh feeding and proportioning of raw materials

(iv) System of introducing controlled amount of air at regular intervals of length in such a way that it does not oxidise the reduced product in the bed,

(v) System of temperature sensing at regular intervals of length of the reactor and recording of the same.

(vi) System of indirect cooling of sponge iron-char mixture in a rotary steel cylindrical shell using water from the outside.

(vii) System of treating waste gases and maintaining desired flow profile through pressure control.

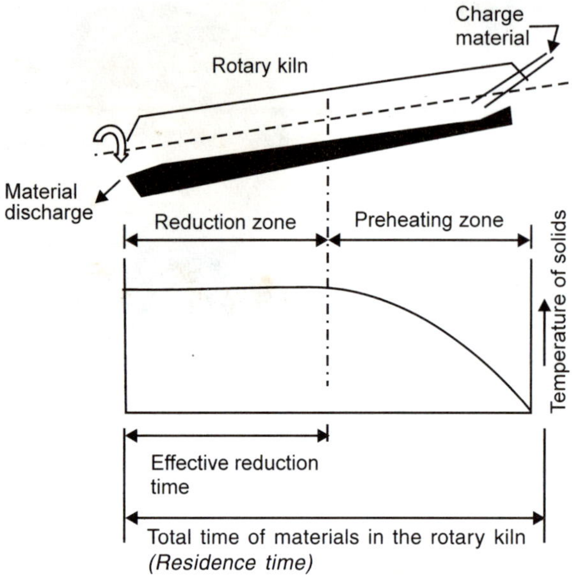

Fig. 2.1 Key steps in sponge iron making in rotary kiln

A typical process scheme for making sponge iron in a rotary kiln is presented in Fig. 2.1. While Fig. 2.1 shows only the key steps, a more detailed scheme, as it would appear for a typical operating plant, is presented in Fig. 2.2.

2.3 SPONGE IRON PILOT PLANT OF RDCIS SAIL

The sponge iron Pilot Plant (SIPP) of RDCIS, SAIL, which would be mentioned a number of times, was set up in 1980-82 with almost all the features of a commercial rotary kiln sponge iron plant. The know-how status was, however, slightly different at that time. Figure 2.3 represents more appropriately the SIPP of RDCIS SAIL.

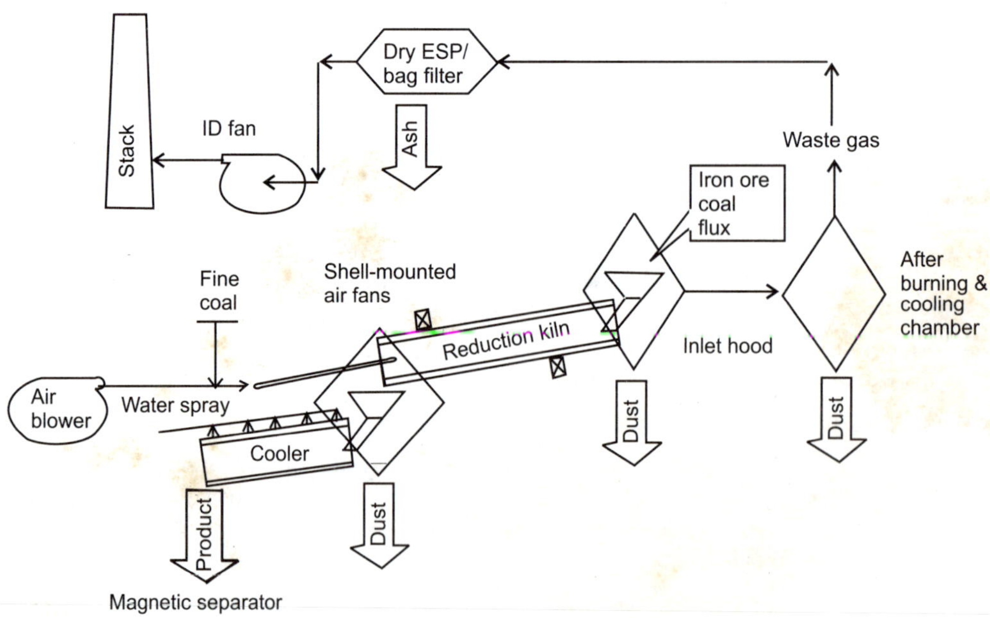

Fig. 2.2 A concise schematic representation of a rotary kiln sponge iron plant

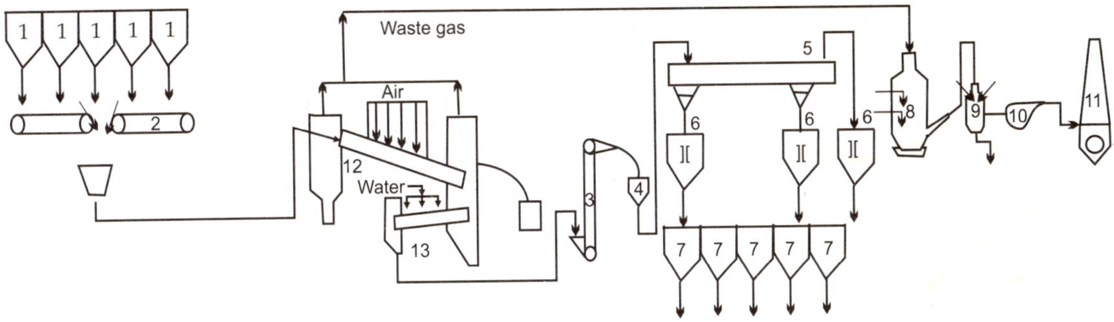

1. Raw material bins
2. Belt conveyor
3. Bucket elevator
4. Surge bin
5. Vibratory screen
6. Magnetic separator
7. Product storage bins
8. After burning chamber
9. Radial flow scrubber
10. Induced draught fan
11. Waste gas stack
12. Rotary kiln
13. Cooler

Fig. 2.3 A schematic of the sponge iron Pilot Plant of RDCIS, SAIL

This Plant of capacity 5 to 9 tonnes per day of sponge iron was commissioned in March 1982 within the premises of M/s HEC at Ranchi, with the objective of adapting and assimilating coal based sponge iron technology in India. The Pilot Plant was in regular operation since its commissioning till 1992-93, with 4 to 5 Campaigns each year.

48 campaigns were carried out in the Pilot Plant with various ore and coal combinations from different deposits in the country. The two longest campaigns lasted 62 days each. A total of 26 ore-coal combinations were processed in the Pilot Plant. Most of the iron ores tested in Pilot Plant were found suitable for sponge iron making. When operated with a good quality coal, metallisation level was consistently above 90%. On the other hand three of the coals tested in Pilot Plant were either not suitable or were only marginally acceptable. In such cases obviously metallisation levels were reduced – in extreme cases upto 70%. Other major results include:

- Development of ore-coal composite pellets technology, which improves kiln productivity and reduces energy consumption (patented)
- Pre-heating system of ore (patented)
- Simultaneous injection of under-bed hydrocarbon fuel and over-bed air in a rotary reactor

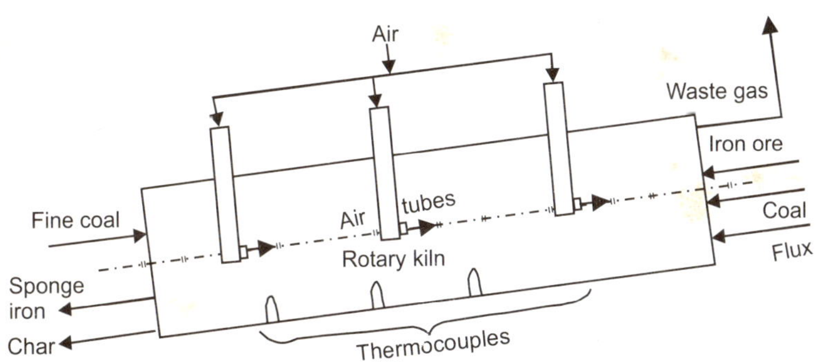

Fig. 2.4 Essential features of a reduction kiln in a rotary kiln sponge iron plant

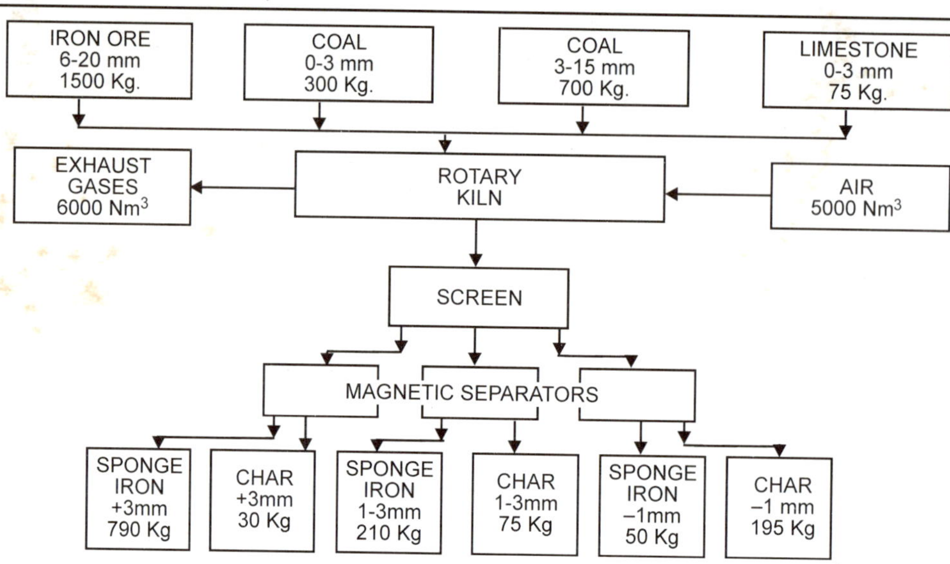

Fig. 2.5 Material balance in a rotary kiln sponge iron plant

2.4 FEATURES OF A ROTARY KILN SPONGE IRON PLANT

Figure 2.4 indicates the essential features which are needed in the reduction kiln of a sponge iron rotary kiln plant. However, the air tubes indicated in this diagram can be substituted by a ported kiln design, which is discussed later in this book.

A typical material balance for the sponge iron making process is presented in Fig. 2.5. Here coal is assumed to contain about 20% ash, something which is hardly available now a days. Product is shown to be screened into three fractions. But due to difficulty in screening the -1 mm fraction, it is usual now a days not to separate out this fraction. Use of 6 to 20 mm iron ore is indicated. Presently it is more common to use 5 to 18 mm fraction. A typical energy balance in the form of Sankey diagram is presented in Fig. 2.6.

Rotary kiln processes have had to compete with gas-based processes. Gas based processes use relatively costlier input such as pellets and reformed natural gas and conversion cost at similar capacities are higher. But even then, gas-based processes have generally found favour due to better and more consistent quality, lower energy consumption and higher module size. It was realised early that rotary kiln processes can be up-scaled only to a limited extent and bigger module size does not mean a higher economy of scale. Modules bigger than 500 tpd were continuously plagued by problems of accretion formation and were maintenance prone. All such modules have now been phased out.

India has been the largest producer of sponge iron since last few years. Its contribution to world DRI production is in excess of 25% at present. India has been increasing the gap with the next country Venezuela. Trends indicate that this gap would continue to increase in the foreseeable future. Iran is third on the line and Mexico, who was once the world leader, is presently placed fourth. One thing may be noted though

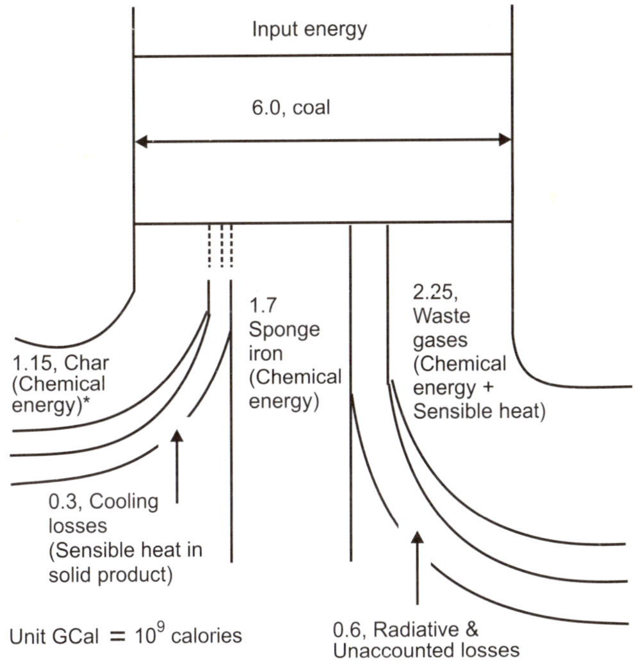

Fig. 2.6 Energy balance in a conventional rotary kiln sponge iron plant

that the entire production by number 2, 3 and 4 countries is through gas based route while in India, more than half the production is by the coal based route.

2.5 THE INDIAN SCENE

Situation in India is different from the rest of the world. Local conditions here have favoured coal based rotary kiln units and presently India has more such modules than rest of the world put together. Over 300 modules are presently in production in India. And there is another trend in India of downscaling. Only in India it has been found profitable to operate 100 tpd and 50 tpd module (even 25 tpd modules), while elsewhere 250 tpd module is considered as the minimum economic size. Table 2.1 and Fig. 2.7 present the production figures of sponge iron in India vis-à-vis world, over the years. Figure 2.8 presents the scene with respect to coal and gas based processes.

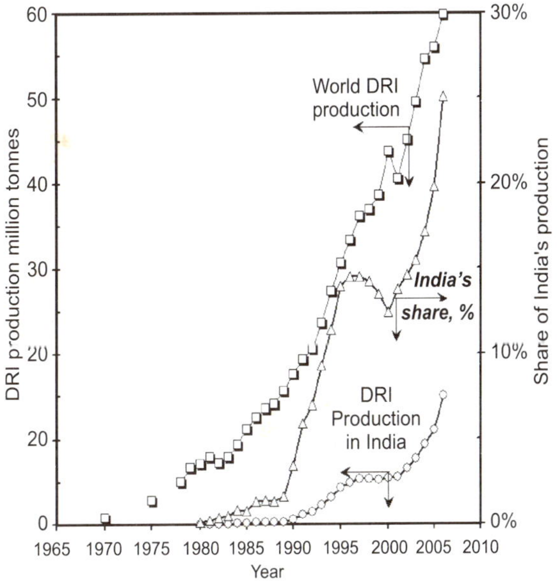

Fig. 2.7 A comparison of sponge iron production of India and World over the years

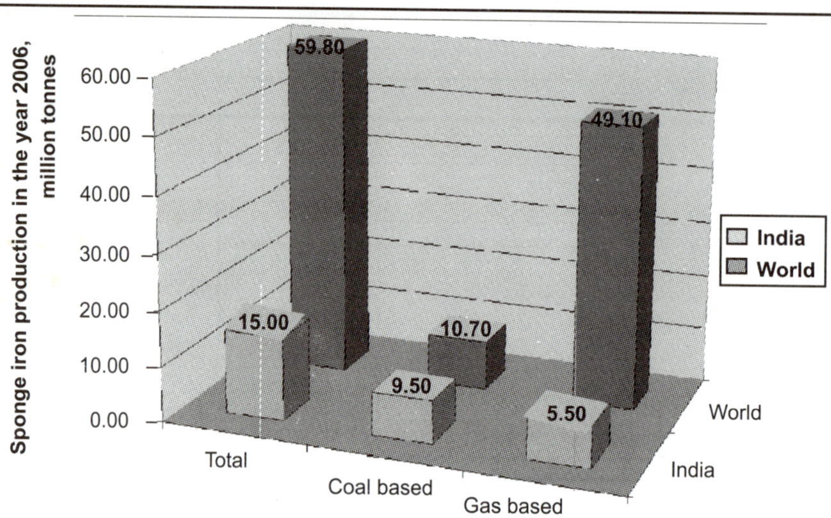

Fig. 2.8 Another comparison of sponge iron production of India and World indicating the dominant position of coal based route in India

The Krupp-Renn process, about which we mentioned in chapter 1, was probably the last of the rotary kiln processes, which attempted to produce iron in fused or semi-fused mass. All of the current processes attempt to prevent any type of fusion during production.

The processes, which are currently in vogue, are Jindal Process (50, 350 and 500 tpd), SIIL Process (100 tpd), OSIL Process (300 and 500 tpd), SL/RN (Lurgi) Process (100 and 500 tpd), Krupp-Codir Process (400 and 500 tpd), DRC Process (250 to 350 tpd) and TDR Process (400 tpd). OSIL Process has evolved from the ported kiln concept of ACCAR Process, while SIIL Process has been based on the Lurgi or SL/RN Process. Presently, of the total sponge iron produced in the world, coal based rotary kiln processes contribute only about 15%, but considering India alone this percentage is about 65%. Trends point to substantial increase in the latter, in spite of the fact that a 2 mtpa gas based module has been commissioned recently in western India.

As mentioned earlier, nearly 35% of India's sponge iron production is accounted for by the three gas based plants located near the western coast. M/s Essar Steel Limited located in Hazira in the state of Gujarat has five operating modules with a total capacity of 5.5 mtpy and claim to be the largest sponge iron plant in the world. Ispat Industries Limited and Vikram Ispat Limited both located in the state of Maharashtra have one module each of capacities 1.4 and 0.9 mtpy.

On the coal based front, the plant of M/s Jindal Steel and Power Limited (JSPL) in Raigarh in the central Indian state of Chhattisgarh is the largest in India and probably in the world. It has ten modules totalling 1.2 mtpy capacity.

Table 2.1 DRI Production: India and World*

Million Tonnes

Year	India	World
1970	0	0.79
1975	0	2.81
1978	0	5.00
1979	0	6.64
1980	0.01	7.14
1981	0.02	7.92
1982	0.03	7.28
1983	0.04	7.90
1984	0.08	9.34
1985	0.09	11.17
1986	0.17	12.53
1987	0.19	13.52
1988	0.19	14.09
1989	0.26	15.63
1990	0.61	17.68
1991	1.15	19.32
1992	1.44	20.51
1993	2.21	23.65
1994	3.12	27.37
1995	4.28	30.67
1996	4.84	33.30
1997	5.26	36.19
1998	5.26	36.96
1999	5.22	38.59
2000	5.44	43.78
2001	5.59	40.51
2002	6.59	45.10
2003	7.67	49.45
2004	9.37	54.60
2005	11.10	55.96
2006	15.00	59.80

*Data taken mainly direct from Midrex

Thus we see that the largest coal based plant in the world barely stands up to the smallest gas based module in India. But then the coal based plants make it up in numbers. As mentioned earlier, over 300 modules of coal based plants are operating in India. And situation is changing so fast, almost on daily basis, that not much point is served by describing these plants here.

Apart from JSPL, some of the plants which operate large size modules (300 to 500 tpd) are Bihar Sponge Iron in Jharkhand, Prakash Industries, Nova Iron and Steel, Monnet Ispat, Godavari Ispat and Power all in Chhattisgarh, Sunflag Iron and Steel Company, Lloyds Metals and Engineers, in Maharashtra, Orissa Sponge Iron, Tata Sponge Iron, in Orissa, GSAL India in Andhra Pradesh, etc. We must make special mention of Sponge Iron India Limited in Paloncha in Andhra Pradesh, which was the first of the commercial plants (originally called a demonstration plant) commissioned in 1980, and which started the race for the installation of the 100 tpd modules, the total number now being well over one hundred. 50 tpd modules could also have exceeded 100 in number. But another special mention must be made of the 25 tpd modules. One is operating in Ramgarh in Jharkhand (Palash Sponge Iron), while at least two modules are operating in Raipur in Chhattisgarh.

2.6 WHY SHOULD WE SELECT A ROTARY KILN?

The rotary kiln direct reduction (RKDR) processes have been looked upon with apprehension, mainly because there have been rapid births and deaths of processes in this group. But the fact that it has re-emerged,

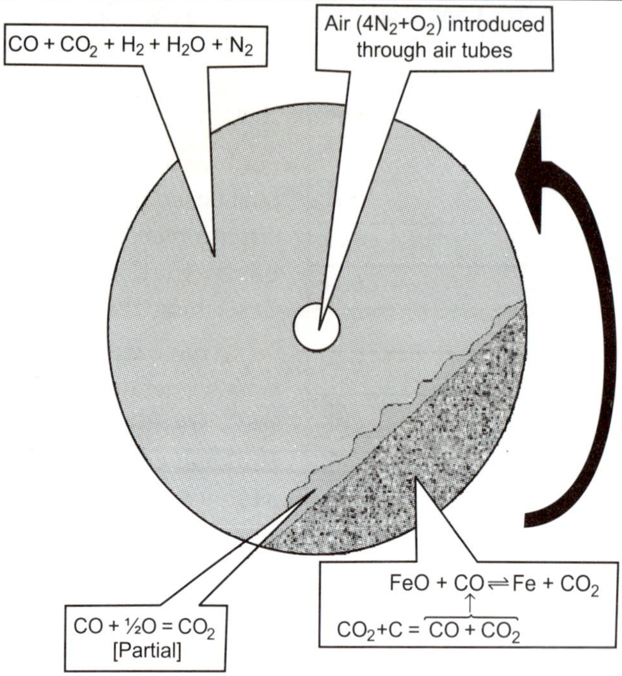

Fig. 2.9 Delicate balance of oxidising and reducing conditions in a sponge iron rotary kiln

points to certain strengths of this process. Let us examine some of them.

2.6.1 Process Strengths

Rotary kiln process has to compete mainly with the shaft process of making sponge iron and in some cases with iron making blast furnace. As compared to them, the rotary kiln has some advantages, as also some limitations, both with respect to the process and the product it makes. The major process strengths of rotary kiln are:

(i) A rotary kiln can mix the solid charge as it heats and reduces it. Simultaneous mixing helps in the dilution of CO_2 concentration formed around the iron ore/sponge iron particles – which is necessary for the reduction reaction to proceed.

(ii) As a large freeboard volume is available above the solid charge (about 85%), the rotary kiln can tolerate heavily dust-laden gas. When the kiln is suitably designed, it would be best suited for utilising the Indian high ash non-cooking coals. In shaft reactors, generation of such dust leads to choking and channelling which leads finally to disruption of the process.

(iii) Rotary kiln can serve the dual purpose of a coal gasifier as well as an ore reducer. Preparation of reducing gas from coal is an expensive step, which is coming in the way of commercialisation of coal gasification based DR process. Therefore, rotary kiln DR process has proved commercially viable, even with low productivity per unit volume, because of this capability to perform two different functions simultaneously.

Figure 2.9 schematically represents the situation inside a rotary kiln where a delicate balance of reducing zone within the chargebed and an oxidising zone in the freeboard is always maintained.

(iv) In comparison to blast furnace, the temperature of reduction of iron oxide is much lower in rotary kiln (about 1000°C as against 1500 to 2000°C in blast furnace). This means that much less energy is required for bringing the reactants to the temperature of reaction.

2.6.2 Product Strengths

Additionally the strengths of the product made by rotary kiln are:

(i) It is easy to desulphurise iron ore while making sponge iron. Consequently the sponge iron of much lower sulphur content can be produced as compared to blast furnace hot metal. For shaft process of sponge iron making, prior and meticulous de-sulphurisation of natural gas is necessary to prevent poisoning of catalyst used for reforming.

(ii) Sponge iron produced from rotary kiln is obtained in close granular size range. This permits charging in electric or other steel making furnaces in a continuous manner, obviating the need for opening and closing of roof. Continuous charging permits partial refining during melting stage as the particle passes through the slag layer into the mixed layer. If adequate melting energy is available, refining time, and consequently, operation time can be considerably reduced.

2.6.3 Weaknesses of the Process

Notwithstanding the above, rotary kiln has a number of weaknesses. These are coming in the way of its wide acceptability. The main process related weaknesses of rotary kiln are:

(i) It has very low productivity. Shaft furnaces, which make sponge iron, give upto five times more output than rotary kilns of same inner volume. Productivity in rotary kiln is consequently much lower.

(ii) The rotating reactor makes it difficult to incorporate process control and quality control systems. Energy saving measures, such as use of pre-heated air, are difficult to incorporate. To prevent ingress of atmospheric air an elaborate sealing system is required, which has made the reactor very "engineering intensive".

(iii) The RKDR process has low energy efficiency. The stored energy in sponge iron is about 1.7 GCal per tonne, while energy usually spent in making it in rotary kiln is about 6 GCal per tonne. Among other things, a lot of energy goes out in waste gases (over 2 GCal per tonne).

(iv) The RKDR process produces some sponge iron in fine form (-3 mm) which is a little difficult to utilise in electric furnaces. While much of the fines are generated due to the nature of ore used, the situation is aggravated by the tumcbling action within the rotary kiln, which forces softer particles to break down further.

2.6.4 Weaknesses of the Product

In addition the sponge iron made by rotary kiln has the following weaknesses:

(i) For charging in electric furnaces in substantial quantities, a system of continuous charging needs to be installed. This would mean an additional investment for the existing units, which are not having this facility.

(ii) The sponge iron from rotary kiln has much lower carbon content (usually 0.2%) than either the sponge iron from shaft furnace (0.7 to 2%) or the hot metal from blast furnace. Carbon in sponge iron not only helps in adding to the opening carbon in molten bath, it also carries in chemical energy, which helps in reducing the consumption of electric power. Too low a carbon content comes in the way of a healthy carbon boil and, therefore, bath carburisers need to be added. Clean carburisers are costly while coke, char or pig iron carries with it undesirable elements like sulphur and phosphorous.

(iii) Sponge iron from rotary kiln carries with it more gangue and phosphorous than those from shaft furnace, mainly because shaft furnace uses cleaner inputs. Gangue and phosphorous contents are much higher than they are in iron and steel scrap, which means extra inputs of phosphorous and slag in electric furnaces.

(iv) When we compare with scrap and pig iron, all sponge irons are prone to re-oxidation and the product from rotary kiln is no exception. However, this rotary kiln sponge iron is much less susceptible to re-oxidation as compared to sponge iron from shaft units using reformed gases.

Those who have ventured into sponge iron have to endeavour to exploit the strengths of RKDR to the fullest extent and would have to try to mitigate the effects of its weaknesses suitably. Those who contemplate venturing into sponge iron have to make a thorough analysis as to whether the strengths outweigh the disadvantages or not in the scenario they are finding themselves in. It becomes the duty of the process developers to put in innovations, which make greater use of the strengths and minimise to the extent possible the weaknesses of RKDR.

There are many basic aspects, which need to be considered for making sponge iron in rotary kiln, the important ones being:

(i) Thermodynamics of reduction and gasification reactions

(ii) Characteristics of raw materials and their role in the process

(iii) Reaction kinetics, roles of reducibility of iron ore and reactivity of coal char and thereby the basis of selection of iron ore and coal

(iv) Movement of solids in the rotary kiln and its residence time

(v) Gas evolution and flow rate

(vi) Heat transfer, temperature profile and process model

These would be dealt in the subsequent chapters.

3 CHAPTER

Thermodynamic Considerations: Feasibility of Reaction

Thermodynamics, and more specifically the free energies of chemical reactions and the equilibrium constants derived therefrom, govern the limits of all chemical and metallurgical processes.

Readers may like to recall that the science of thermodynamics evolved out of the study of those properties of heat which are not found in other forms of energy. Heat flows much like a liquid seeking its own level (i.e., stabilisation of temperature). While mechanical work can be completely converted into heat through, for example, friction, conversion of heat into work can only be achieved when accompanied with simultaneous discard of a certain lower quality (lower temperature) heat. During the most efficient operation of such a heat engine, the heat per unit absolute temperature (or the entropy) remains constant. But as the most efficient operation takes infinite time, such an operation cannot be a practical operation. For actual operation, entropy increases. For a proposed operation or reaction, an indicated increase in entropy signifies its feasibility (i.e., the process or operation has a tendency to move into the forward direction). For isothermal reactions, free energy (negative of entropy multiplied by absolute temperature) is a more convenient term for expressing entropy. It is in energy units and can be conceived as the total energy less the bound energy. Feasible reactions are indicated by a decrease in free energy. And the extent of decrease in free energy is the driving force of the reaction.

Thermodynamics and free energy change assume even greater importance in processes involving solids, as in sponge iron making. Sponge iron making gave way to a more indirect process route involving melting and is now being resurrected only after a greater understanding of how to provide a greater driving force so that reaction rates reach such levels, which can compete with indirect processes.

Figure 3.1 is an approximate representation of Ellingham diagram of a few key oxides of our interest. For a more accurate representation the publications of United States Steel or of the Association of Iron and Steel Engineers may be referred to.

An Ellingham diagram plots the standard free energy changes of reactions and their variation with temperature. The term 'standard' refers to a state of reference, generally 1 atmosphere pressure and a temperatue of either 0°C (273°K) or, as in the present case, 25°C (298°K) and sometimes 1000°C (1273°K). In case of Ellingham diagram for oxides, which is relevant to our subject, the lines lying higher up represent a less stable oxide as compared to the lines lower down. It can be seen that the line of CO_2 formation from CO and O_2 lies slightly above the line of FeO formation from iron and oxygen. This means that under standard state the following reaction cannot proceed forward.

$$FeO + CO \leftrightarrow Fe + CO_2$$

This is the key reaction in sponge iron making process and the reaction is made to move in the forward direction by keeping a large excess of CO over CO_2; i.e., by heavily reducing the activity of CO_2.

Table 3.1 presents the thermodynamic parameters that are relevant to sponge iron making in rotary kiln. In this Table, the symbols G and H represents free energy and heat content (enthalpy) respectively. Superscript 'o' indicates standard state and subscripts '298' and '1273' indicate the temperature in degreees Kelvin (°K). From the Table we can deduce the following:

(a) The overall reduction process is highly endothermic and additional coal (or any other fuel) is required to be combusted to meet this heat demand.

(b) As stated earlier, under standard conditions and 1000°C, reduction of FeO to Fe is not feasible (reaction no. 2 in the Table). But it becomes feasible if CO/CO_2 ratio is maintained above 2.52; i.e., the concentration of CO_2 in the vicinity of iron ore particles is not allowed to build up above 28%.

(c) The above is achieved by maintaining a fast rate of the gasification reaction (reaction no.6). If the reduction is carried out at a lower temperature,

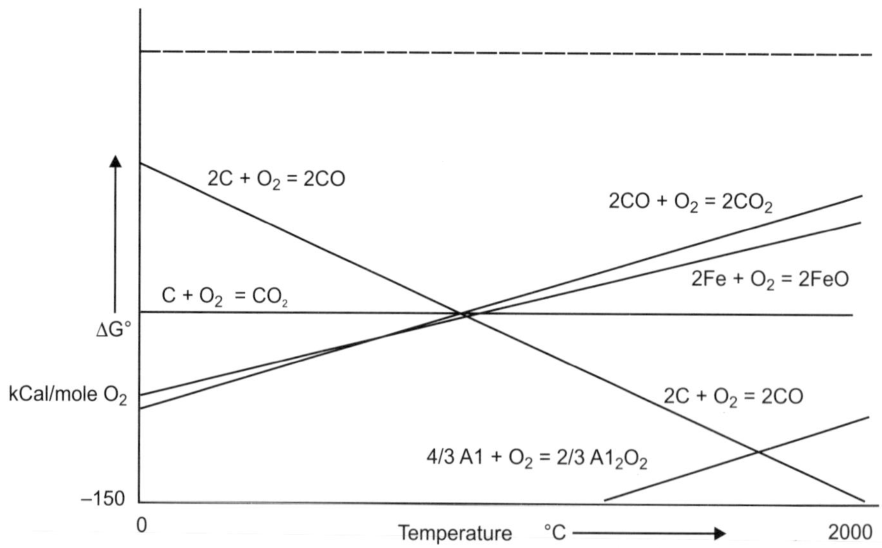

Fig. 3.1 Ellingham diagram for a few key oxides

Table 3.1 Thermodynamic data of important reactions

ΔG and ΔH values in kCal/kg mole, T in °K

SL.	Class of reaction	Reaction	ΔG^o_T	ΔG^o_{298}	ΔG^o_{1273}	ΔH^o_{298}	ΔH^o_{1273}	Remarks
1.	Reduction reaction	$Fe_2O_3 + CO = 2FeO + CO_2$	+ 2,120 – 10.39 T	– 976	–11,106	+ 2,270	– 2,532	Always feasible. Easily ccomplished at the start of reduction zone of rotary kiln.
2.		$FeO + CO = Fe + CO_2$	– 4,190 + 5.13 T	– 2,661	+ 2,340	– 4,430	– 8,530	Key reaction. CO/CO_2 ratio required to be over 2.52 at 1000°C for reaction to proceed
3.	Gasification and Combustion reaction	$C + ½ O_2 = CO$	– 26,700–20.95 T	– 32,943	– 53,369	– 26,420	– 27,183	Combuction (Heat generating) reactions
4.		$C + O_2 = CO_2$	– 94,200 – 0.2 T	– 94,260	– 94,455	– 94,050	– 94,561	
5.		$C + ½ O_2 = CO_2$	– 67,500+20.75 T	– 61,316	– 41,085	– 67,630	– 67,378	
6.		$CO_2 + C = 2CO$	+ 40,800 – 41.70 T	+ 28,373	– 12,284	+ 41,210	+ 40,195	Gasification of carbon
7.	Overall reaction	$FeO + C = Fe + CO$	+ 36,610 – 36.57 T	+ 25,712	– 9,944	+ 36,780	+ 31,665	Over all reaction for the reduction zone.
8.		$Fe_2O_3 + 3/2\ C = 2Fe + 3/2\ CO_2$	+ 54,940 – 62.68 T	+ 36,261	– 24,852	+ 55,225	+ 40,700	Over all reaction for the RKDR process. The highly endothermic heat requirement is met by burning of additional coal.

Data taken or calculated from Kubaschewski and Alcock – *Metallurgical Thermochemistry*, Pergamon Press, 1979

higher percentages of CO_2 can be tolerated.

While gasification (reaction no. 6) is feasible at temperature of operation of rotary kiln, the rate of gasification is critically dependent on the nature of surface of char (as also the interior, since surface is continuously consumed) and thereby the nature of coal used. A coal, which yields reactive char, which can quickly convert traces of CO_2 to CO, is necessary to make the process move forward. This has further been discussed under Reactivity of Coal char in the chapter on Raw Materials (Chapter 9).

Kindly note that FeO is a non-stoichiometric compound. Extent of non-stoichiometry has been ignored and there could be a very marginal error in the data on this account. Non-stoichiometry means that atom ratio of Fe and O is not exactly 1:1 but slightly different (roughly about 0.95:1).

4 CHAPTER

Aerodynamics Inside a Sponge Iron Rotary Kiln

Aerodynamics refers to the science of study of gas flow and the forces applied by the flowing gas on interacting bodies. The flow pattern of gases inside a sponge iron rotary kiln has a strong influence on the satisfactory performance of the process. One important aspect, which the readers would have guessed, is the need to maintain a delicate balance of reducing atmosphere near the surface of solid charge in the reduction zone and oxidising atmosphere at and near the axis of the kiln. This and a few other aspects are important in the study of gas flow characteristics inside a sponge iron rotary kiln.

4.1 THUMB RULES USED IN PRODUCTION OF SPONGE IRON

Before discussing the aerodynamics or the gas flow characteristics inside a sponge iron rotary kiln, we would like to discuss a few thumb-rules being followed by rotary kiln sponge iron manufacturers, which have direct linkage to gas flow. These thumb-rules are results of experience gained in a long history of empirical and industrial practice.

Sponge iron manufacturers using coal alone as the reducing and heat supply agent, have not been able to produce satisfactory sponge iron without the use of 'injection coal' i.e., coal thrown or slung from the discharge end of the rotary kiln. A large portion of the thrown coal falls on the solid bed close to the discharge end and passes on to cooler before it is fully de-volatilised. Thus de-volatilisation of coal is continued in the early part of the cooler. As a thumb-rule, for the protection of the sponge iron produced, a ratio of fixed carbon to total iron in the kiln discharge product is maintained at 0.15. This signifies a source of loss of carbon from the process going unutilised. But, for the sake of obtaining a satisfactory product at the end, producers normally do not attempt to reduce this ratio.

Why so much carbon is required to be kept in the product? In fact if the atmosphere in the cooler is fully reducing i.e., CO_2, H_2O and O_2 are totally absent, there is no need to keep that much carbon in the product. But as discussed earlier, due to the need to keep a delicate balance between oxidising and reducing conditions within the free-board in the kiln, there is likelihood of some CO_2 and H_2O diffusing out from kiln into the cooler atmosphere. Further, there may be some air ingress, if the system is not perfectly sealed (or there is otherwise fault in or wearing out of the sealing system). Hence, there is need to protect the sponge iron from re-oxidation in the cooler, until its temperature drops below its ignition point (Crossing Point—kindly refer Chapter 11).

But fixed carbon in the char is ill equipped to protect the hot sponge iron. It is not the fixed carbon, but the little bit of volatile matter that keeps liberating, which helps in protecting the sponge iron.

The above discussion was needed before we discuss Aerodynamics, in order to emphasise the point that at kiln discharge, the gas velocity in the free board is not zero and some gas evolved in cooler joins the rotary kiln gases.

4.2 SOURCES OF GAS

Apart from the above, and the air introduced in rotary kiln for combustion purposes, almost all the gas in rotary kiln is generated from within the solid bed. Main reactions taking place within the bed are

$$FeO + CO = Fe + CO_2$$
$$CO_2 + C = 2CO$$

Overall reaction within the bed becomes (by adding the above two reactions)

$$FeO + C = Fe + CO$$

Thus there is surplus of CO in the bed which keeps evolving as the iron oxide is being metallised. Even in the region where Fe_2O_3 is being converted to FeO by the reaction,

$$Fe_2O_3 + CO = 2FeO + CO_2$$
$$CO_2 + C = 2CO$$

The overall reaction being,

$$Fe_2O_3 + C = 2FeO + CO$$

There is still a surplus of CO, which keeps evolving from the bed. Thus from the point at which solid bed has reached about 700°C till its discharge after full reduction at a temperature close to 1000°C, CO gas continuously evolves. Before this point, at first moisture of the total charge, and then volatile matter from feed coal keeps evolving out from bed. Even the coal thrown in pneumatically from discharge end, which falls on solid bed in the reduction zone only, evolves moisture and VM from the solid bed.

Thus gases evolve almost uniformly from the solid bed throughout the kiln length and the sum total comes out from the charging end of the kiln. The gas volume is supplemented by the controlled amount of air injected through the Central Burner (a misnomer, since it is not a burner and only injects air and the Shell Air Fans (mounted on the kiln shell) via the Air Tubes, placed all along the length of the kiln.

4.3 EFFECTS OF AIR INJECTION

Air can be taken to be consisting of one part oxygen and four parts nitrogen. Nitrogen adds to the gas volume while oxygen is consumed without any addition to the gas volume.

$$2CO + O_2 = 2CO_2$$
$$2H_2 + O_2 = 2H_2O$$

Thus injecting air results in increasing gas volume by 80% of the volume of air injected. Air is injected in controlled quantities for combusting CO (& other combustibles) in free board to maintain the desired temperature profile and meet the thermal needs of the process.

Thus gas velocity keeps increasing from nearly nil at kiln discharge end to the maximum level at charging end of kiln (or at the After Burning and Cooling – ABC chamber).

4.4 LIMITS TO GAS VELOCITY

A very high velocity of gas is expected to blow away some of the granular solids from the bed, which is obviously not desired. This is expected to happen at fluidising velocity (velocity of gas at which the suspended granular particles remain in suspension and do not settle down – the suspended solids thus start behaving like a fluid). As a thumb-rule a velocity value of 60m/sec is considered as the fluidising (and therefore the limiting) velocity. Fortunately, due to the need to maintain a large free board in the reduction zone of the kiln, designs of conventional sponge iron rotary kilns are not constrained by this consideration. However, as and when the need arises to evolve more efficient kiln designs, fluidising velocity considerations will have to be kept in mind.

4.5 GAS FLOW PATTERN

In order to keep the oxidising constituents existing at and near the kiln axis in the reduction zone, to remain mostly separate from the surrounding gases, it is important that the gas flow in this region be lamellar; i.e., the layer of oxidising gases flow as a separate layer and do not mix with the layers of reducing gases surrounding it. In other words, in the reduction zone the flow should not be turbulent, meaning small layers (eddy's) of gases do not form and mix laterally with the reducing gases in the vicinity.

On the other hand if turbulence were not created in the preheating zone, heat transfer would be heavily impaired. So, while a lamellar flow pattern is desirable in the reduction zone of the kiln, turbulent flow pattern is preferred in the preheating zone. Based on extended investigations on the flow of fluids in pipes, Osborne Reynold in 1883 introduced the empirical criterion of a dimensionless number, which later came to be known as the Reynold's Number, in determining whether a particular flow would be lamellar or turbulent.

Reynold's Number N_R is defined by the following expression

$$N_R = DV\rho/\mu$$

where D = Diameter of pipe

Here, diameter of rotary kiln would be applicable

V = Velocity of the fluid

Here, gas velocity in the kiln

ρ = Density of fluid (kiln gas)

μ = Viscosity of fluid (kiln gas)

Typically, if the value of N_R is below 2,000, the flow would be lamellar, and above 4,000 the flow would have turbulent pattern. Between the values of 2,000 to 4,000 the flow pattern would be of mixed type, in some regions lamellar and in other regions turbulent. If care is taken to limit the filling degree to within 15% in the reduction zone, values of Reynold's Number are then maintained within 2000 in some part of reduction zone, close to the discharge point of solids, and flow pattern remains lamellar. Beyond this point, towards the charging end of solids, and in the entire preheating zone, gas volumes increase rapidly and the flow pattern then reaches mixed zone and thereafter turbulent zone and thereby heat transfer operation is facilitated.

4.6 CFD ANALYSIS

As mentioned earlier, making sponge iron in rotary kiln using coal is a delicate balancing

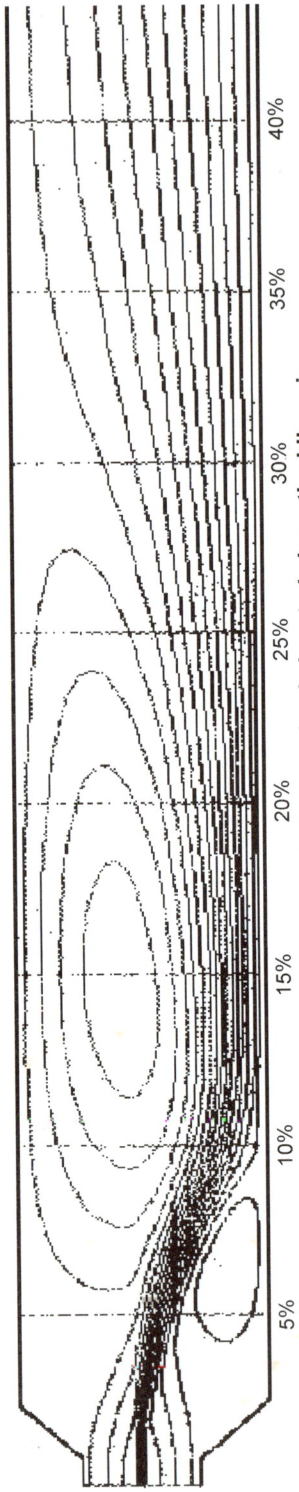

Fig. 4.1 Gas flow profile when air injection pipe is located along the kiln axis

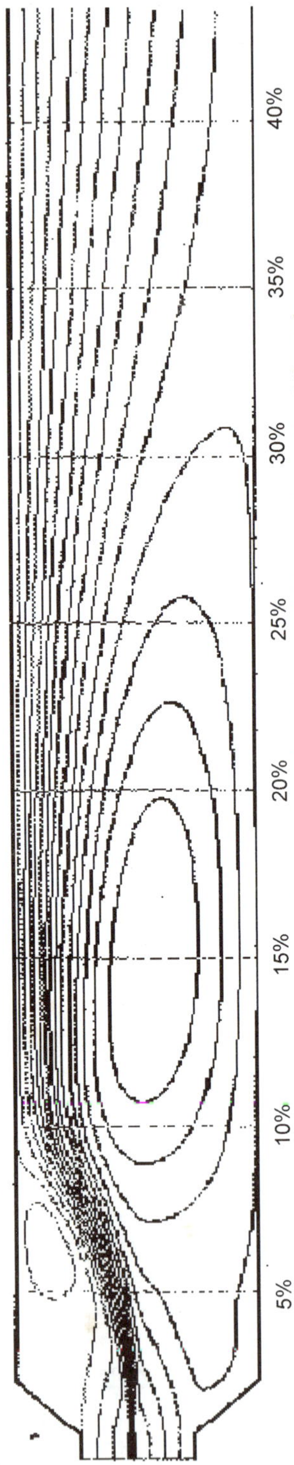

Fig. 4.2 Gas flow profile when air injection pipe is located 50 mm above the kiln axis

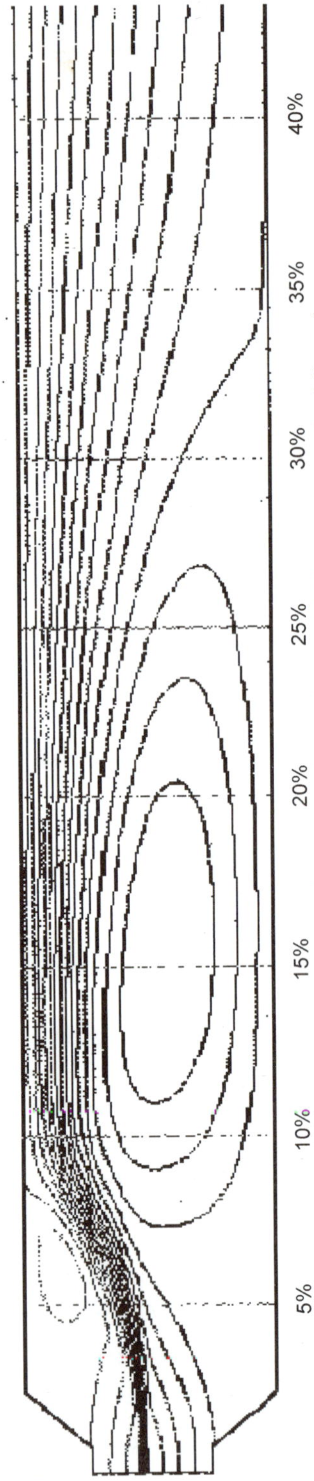

Fig. 4.3 Gas flow profile when air injection pipe is located 100 mm above the kiln axis

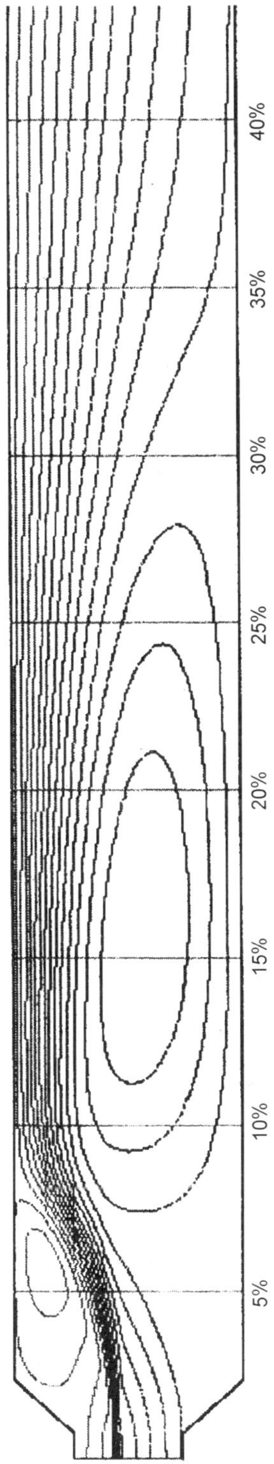

Fig. 4.4 Gas flow profile when air injection pipe is located 150 mm above the kiln axis

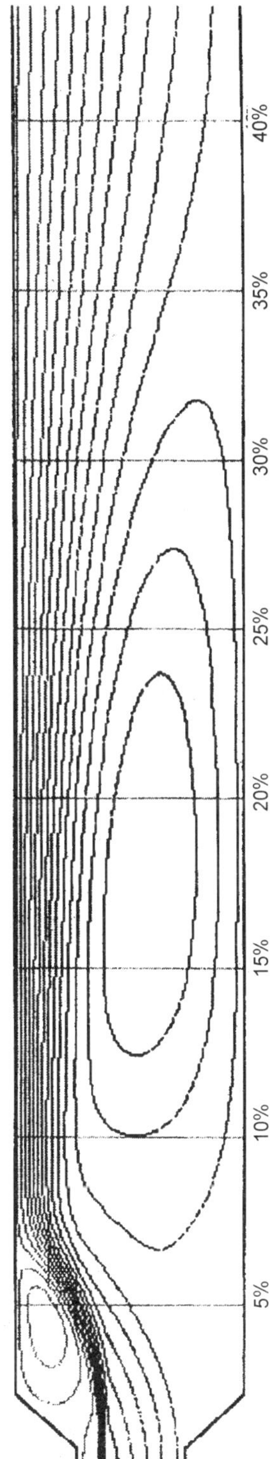

Fig. 4.5 Gas flow profile when air injection pipe is located 320 mm above the kiln axis

exercise. In the same cross section in reduction zone we need to maintain fully reducing atmosphere within, and just above, the solid charge bed, and close to the axis of the kiln we need to oxidise the gases to generate heat and also keep fair percentages of trivalent molecules H_2O and CO_2 so that adequate heat transfer by radiation is achieved. Readers may note here that bivalent gases like CO, H_2, N_2 radiate negligible amount of heat. And this oxidation of CO to CO_2 and H_2 to H_2O should be carried out in a manner such that the reducing atmosphere within, and just above, the solid charge is not disturbed. To achieve this, it has been found necessary to maintain a large free board. Consequently, it has not been found practicable to maintain average filling degree of solids of more than 15% in the reduction zone. Sometimes, even this 15% filling is found to be too large as the bed atmosphere is found to get disturbed by the flow pattern in the free board. The result is a lower metallised or a re-oxidised sponge iron. In extreme cases the heat generated in re-oxidation makes the sponge iron pieces to fuse together and form agglomerates.

It is possible to determine the expected flow pattern inside a sponge iron rotary kiln using Computational Fluid Dynamics or the CFD technique. Fluid flow is a highly complex process. Fluid characteristics, specially its viscosity, keep changing with both pressure and temperature. The flow pattern can be computed by the finite element analysis, where the entire volume is broken into a grid of very small but finite elemental volumes within which the properties can be assumed to be uniform. Based on the input data to a particular elemental volume, and the interactive effect of the elements surrounding it, the output data and the flow direction are computed. The flow direction decides which element is the successor to the first element and the output data of the first element applies to this second element. The computation is thus continued until we reach the last element.

Computational steps and the number of input and intermediate data being extremely large, and the interactive steps of calculation being extremely complex, the solution requires a computer capable of storing and handling such a large number of data, and a computer program, which can handle and interpret such a complex process.

In a typical case, RDCIS, SAIL, used FLUENT software and a personal computer with a Pentium-4 processor to compute flow pattern in a typical 100 tpd sponge iron rotary kiln. The flow pattern generated by the introduction of "Combustion Air" at the kiln discharge end using a 65 mm diameter pipe at the rate of about 140 m/s was studied. The kiln is 42 m long and available inside diameter is 2600 mm. The kiln is conical at discharge end with a 45° cone and discharge comes out through an opening of 1220 mm diameter or 610 mm radius. If the air required (combustion air) is introduced through an axially placed pipe of 65 mm dia. at a linear velocity of about 140 m/s, the CFD analysis indicates an airflow profile represented in Fig. 4.1. From this figure it can be seen that at 4.2 m or 10% of the length of the kiln, the air injected would tend to strike the kiln bottom where the solid bed would be found during sponge iron making operation. This CFD analysis has been carried out in an empty kiln, and can be repeated with the solid bed in place.

The above result predicts the ill effects

of placing the air injection pipe axially, which would disturb the reducing atmosphere within the solid bed quite intensely and thus would come in the way of obtaining a well metallised product.

Figures 4.2, 4.3, 4.4 and 4.5 show the results of CFD analysis when the air pipe is placed 50 mm, 100 mm, 150 mm and 320 mm above the axis. In all these cases it is seen that the layers of air flow tend to bend upwards first and, after what may be called a recoil from the top surface, tends to strike the bottom beyond 15 m or 35% of the length of the kiln. By this time it would have lost most of its momentum, and chemical reactions would have consumed all O_2 in the stream for this strike on the solid bed to have any deleterious effect.

Additionally, there is a vortex like formation below these layers of air stream in Fig. 4.2 to 4.5. A portion of kiln gas tends to re-circulate over the bottom layer (where solid would normally be present in actual operation). This layer, which would come in contact with the solids, would consist of only reducing gases. Even then it is not desirable to disturb the gas dynamics within the solid bed. A rough estimate of the velocity of strike would be between 3 to 4 m/s which is much smaller than the general levels of velocities in the kiln free board. Lowest velocity is indicated when air pipe is placed 320 mm above the axis of the kiln.

The reason why the value of 320 mm was chosen is because some of the sponge iron plants, operating 100 tpd kilns, have claimed to have found best results when the bulk of air introduced from discharge end is injected using a pipe placed 320 mm above the axis of the kiln. Thus, although the optimum location of air pipe has been an exercise of trial and error in the past but Computational Fluid Dynamics or CFD exercise has brought out the logic behind the suitability of a particular location.

Very few variations were attempted in the exercise cited above. A more rigorous application of CFD analysis to the conditions of an actually operating rotary kiln should help in further optimisation of process parameters and thereby further enhancing the productivity and quality of the rotary kiln sponge iron making process.

5 CHAPTER

Mathematical Modelling in Rotary Kiln Sponge Iron Making

Mathematical models are being increasingly used in the understanding and control of industrial processes. Use of Mathematical Model ensures that the process operates in the most optimum regime, where the maximum benefit is realized within the available constraints. Since industrial processes generally operate on a large scale, operating it even marginally away from the optimum results in massive loss of opportunities. On the other hand a mathematical model of an industrial process is highly complex and, if the user is not the maker of the model, one finds it very difficult to understand and implement. Therefore, those marketing commercial mathematical models, offer a training module along with the model. Further, makers of commercial mathematical models make them for a wider application. Fine tuning the same to make it suitable for the needs of the customer has to be done after detailed interaction between the maker and the customer. The actual fine tuning may be done by the maker or the customer or jointly between them.

Mathematical models being highly complex, the potential users generally approach it with a sense of scepticism, apprehension and fright. It is, therefore, highly recommended that the potential users should attempt to develop the mathematical model themselves. If they are successful in their attempt the rewards are highly significant, but even if they are unable to develop a workable model, the experience gained would place them in a far better position to interact with a commercial supplier of a mathematical model than he would otherwise have been.

5.1 WHAT IS A MODEL?

We use a number of models of various types in our day to day life. A model is something which represents or simulates or looks like or even acts like the real thing. We buy for our children a toy car, which is basically meant for them to play with, but it also simultaneously helps the child to understand cars better. Toy car is an example of a physical model which is a smaller version of the real car. Most of the physical models are of this type. Even prototype working models of a machine of newer design is a smaller version of the real thing to come, and is made for the purpose of the validation of the new design. But physical models need not always be smaller than the real thing. A physical model of a DNA molecule is many many times bigger just because it is impossible for the naked eye to physically see objects at molecular scale.

5.2 WHAT IS A MATHEMATICAL MODEL?

These are mathematical expressions, which theoretically or empirically relate parameters of a real world problem to each other. The definition is best understood through examples. We have consciously or unconsciously been using simple mathematical models since ages. Many times we tell a friend over the phone "I shall be with you in 15 minutes." Let us see what we have done. We have analysed the activities required to reach the friend. We have estimated time required for each of the activities and we have added them up. We have also added a few minutes for contingencies and thus have arrived at the final figure.

Ancient Egyptians were experts at making pyramids. They wanted to cut stones at right angles (90°). For this purpose they made three sticks of lengths three units, four units and five units. They joined these sticks at the edges and used the largest angle of the triangle formed for matching the corner angles of the stones. When we see a lightening flash, we can count seconds until we hear the thunder, divide seconds by five and conclude that the cloud that caused this is so many miles away. Also if we stand on a quiet sea shore and we see a boat's mast setting below the horizon, we can conclude that the boat has gone beyond a mile range.

5.3 HOW CAN WE MAKE A USEFUL MATHEMATICAL MODEL?

Formulation of a mathematical model starts with identifying the basic principles and fundamental processes involved and converting them into mathematical terms. This identification process is started with the analysis of inputs and outputs of material and energy – in other words the Material Balance and the Energy Balance of the process. Once the overall material balance and the energy balance are worked out, the material and energy flow within the process is analysed. The total process is broken into interlinked sub-processes and each sub-processes and its relationship with other sub-processes are analysed and converted into mathematical terms until a workable mathematical model is developed. After this first development, the model needs to be validated which is best done by inputting actual operating data and the model output is compared with actual output. Based on the observed difference, a correction or adjustment may need to be incorporated in the model. After full validation, the mathematical model is ready for implementation.

While converting the actual process steps into mathematical expressions, a number of simplifying assumptions may need to be made. It is advisable to keep listing these assumptions. During review and validation, these assumptions may need to be examined for their validity or may need to be modified.

5.4 EXAMPLE OF A SMALL MATHEMATICAL MODEL

Let us attempt to make the mathematical model of a small step in sponge iron making (actually a sub-process of sponge iron making), i.e. how to predict the composition and quantity of sponge iron, having been given the composition of iron ore (or any other iron oxide feed such as iron ore pellets). For this we need the complete analysis of iron ore. Then we list the analysis in a Table of the following type (an Excel or a Lotus 1-2-3 spreadsheet). Here, items like Fe_2O_3 and oxygen are not obtained by chemical analysis, but are calculated based on the values of Fe, Fe(metallic), and FeO. Although the assumption that LOI (Loss on Ignition) is 50% CO_2 and 50% H_2O, will not be needed further in this part of the model, but would be used in other parts of mathematical model of sponge iron making.

It is very easy to fill up the columns next to the composition of iron ore and appears almost a superfluous column. But from here we can fill up the next column where oxygen remains only 5% of the previous column. Moisture and LOI are lost in sponge iron making process and so also Fe_2O_3. Fe(t), $SiO_2 + Al_2O_3$ and phosphorous remain unchanged. Sulphur and carbon values are difficult to fill at this point and we leave these columns blank and add the column up (we get here the value of 69.35 kg.). By experience we know that a well metallised sponge iron is:

Table 5.1 Chart for calculating sponge iron composition from iron ore composition

Sl. No.	*Element/Item*	*Iron ore composition*	*Kg of element/ item in 100 kg of iron ore*	*Kg. of element /item in sponge iron, 95% reduced (92% metallised)*	*% of element / item in sponge iron*
1.	Fe (t)	65.00	65.00		
2.	Fe (m)	0.00	0.00		
3.	FeO	0.00	0.00		
4.	Fe_2O_3	92.90	92.90		
5.	Oxygen	27.90	27.90		
6.	Moisture (H_2O)	2.00	2.00		
7.	Loss on ignition ($CO_2 + H_2O$ = 50:50)	2.00	2.00		
8.	$SiO_2 + Al_2O_3$	2.90	2.90		
9.	S	0.05	0.05		
10.	P	0.05	0.05		
11.	C	0.00	0.00		
	Total	100.00	100.00		
Total without sulphur and carbon					

- about 95% reduced (this fact we have already used in calculation)
- contains about 0.02% sulphur and
- also contains about 0.2% carbon

Thus, 69.35 kg of items in sponge iron should account for 99.78% of the items in sponge iron. Thus by multiplying the figures in the fifth column by a factor of (99.78/69.35) we fill up the last column. And by multiplying the carbon and sulphur figures in the last column by (69.35/99.78) we complete the fifth column.

An important inference obtained from this Table is that from 100 kg iron ore we can obtain 69.50 kg of sponge iron. Therefore, 1500 kg of iron ore would give 1042.5 kg of sponge iron. By practical experience we know that during rotary kiln sponge iron making, for each tonne sponge iron made, about 40 kg. of sponge iron (or its equivalent iron ore) is lost in various dust streams. Thus, for making 1000 kg of saleable sponge iron we need to put in 1500 kg of iron ore.

5.5 ROLE OF COAL IN ROTARY KILN SPONGE IRON MAKING PROCESS

For taking up the work of preparing the mathematical model of the rotary kiln DR

Table 5.2 Filled up chart of Table 5.1

Sl. No.	*Element/Item*	*Iron ore composition, %*	*Kg. of element/ item in 100 kg of iron ore*	*Kg of element/ item in sponge iron 95% reduced*	*% of element / item in sponge iron*
1.	Fe (t)	65.00	65.00	65.00	93.50
2.	Fe (m)	0.00	0.00	60.10	86.50
3.	FeO	0.00	0.00	6.30	9.10
4.	Fe_2O_3	92.90	92.90	0.00	0.00
5.	Oxygen	27.90	27.90	1.40	2.00
6.	Moisture (H_2O)	2.00	2.00	0.00	0.00
7.	Loss on ignition (CO_2 + H_2O = 50:50)	2.00	2.00	0.00	0.00
8.	SiO_2 + Al_2O_3	2.90	2.90	2.90	4.10
9.	S	0.05	0.05	0.05	0.02
10.	P	0.05	0.05	0.05	0.07
11.	C	0.00	0.00	0.10	0.20
Total		**100.00**	**100.00**	**69.50**	**100**
Total without sulphur and carbon				69.35	99.78

process, we have to understand at first the role of coal in this process. Coal has got multiple roles in this process. It acts as the source of heat which is required in the process and this is provided by the controlled combustion of carbon and hydrogen in coal by air along the length of the kiln. Carbon in coal also acts as reductant for removing the oxygen in combination with iron in iron ore. This is a third role of coal which is less understood. Coal in the form of char, which accompanies the sponge iron produced from the kiln, helps in protecting the sponge iron from re-oxidation.

Experience has indicated that a ratio of Fixed Carbon to Total Iron in the product should be about 0.15 for a smooth operation of the process.

While coal provides so many services in the process, it has some negative roles also to perform. Coal has the responsibility of raising the temperature of furnace and the furnace charge to the processing temperature. But in doing so it must heat itself to the processing temperature first. In doing so it liberates the volatile matter and the heat required to liberate the volatile matter has also to be pre-accounted for by coal, before it can liberate heat for other purposes. The ash within the coal has to be likewise heated to the processing temperature. Air needed for combustion also needs to be brought to processing temperatures. Likewise, the waste gases which leave the reactor at temperatures somewhat close to processing temperature also carries energy obtained from coal itself.

5.6 QUANTITY OF PROTECTIVE CHAR

It would be convenient to calculate at first the quantity of coal needed to provide the necessary carbon in product discharge to bring the level of Fixed Carbon to Total Iron in product to 0.15. First we infer the VM in coal from its ultimate analysis. Then we devolatilise the coal to obtain the quantity and composition of the char. Then we calculate the quantity of this char in the product to obtain FC/VM = 0.15. In practice the char in the product discharge will have a much lesser fixed carbon than the char, which we have assumed here to be with the product. We would further be assuming that this char gets mixed with a further quantity of ash and the result is the quantity and composition obtained in actual practice.

5.7 REDUCTION OF IRON OXIDE

We now assume that the VM liberated in the above step consists of C, H, H_2O and CO_2. Since sulphur largely passes from coal to flux (one solid to another), we consider it non-volatile and eliminate it from calculations. The available carbon and hydrogen in the VM is then consumed for stoichiometrically reducing the iron ore.

$$Fe_2O_3 + 3/2\ C = 2\ Fe + 3/2\ CO_2$$

$$Fe_2O_3 + 3\ H_2 = 2\ Fe + 3\ H_2O$$

The quantity of iron ore reduced in these steps is estimated as also the heat deficit caused in these steps. Next an estimate is made of the coal which is required for reducing the balance of unreduced iron ore. Also the heat deficit caused by these steps is estimated.

5.8 COAL FOR COMBUSTION

The accumulated heat deficit has to be made up by burning additional coal. The available heat from coal is calculated and based on this the quantity of additional coal required is estimated. Please note that of the total calorific value of coal, some part of the heat is used up by the coal for its own requirement, which is provided for in the calculation beforehand.

Table 5.3 Products of heating 100 kg coal

Assumptions:

Coal composition: Free moisture = 5.6%, Combined moisture =1.8%, Ash = 30.0%, C = 51.4%, H = 3.0%, S = 0.5%, N = 1.3%, CO_2 as carbonate = 0.4%, O = 6.0%.

Heat effects on conversion have been neglected

Quantity in kilogrammes where not mentioned

Constituent	Initial quantity	Quantity being converted	Quantity remaining after conversion	Other constituents consumed	Quantity	Gaseous product constituent	Quantity	Overall product		Volatile or Non volatile	Kg-moles of evolved gases	Volume % of constituents in VM	Volume percentage of constituents in the total gases evolved	Char analysis	
								Constituents	Quantity					Quantity	Percent
Ash	30.0	0	30.0	-	0	-	0	Ash	30.0	Non Volatile				30	43.86%
H_2O (Free)	5.6	0	5.6	-	0	-	0	H_2O (Free)	5.6	Volatile	0.311				
H_2O Combined	1.8	0	1.8	-	0	-	0	H_2O Combined	1.8	Volatile	0.100	7.81%			
CO_2	0.4	0	0.4	-	0	-	0	CO_2	0.4	Volatile	0.009	0.71%	0.57%		
N	1.3	1.3	0	-	0	N_2	1.3	N_2	1.3	Volatile	0.046	3.63%	2.92%		
S	0.5	0	0.5	-	0	-	0	S*	0.5	Non Volatile				0.5	0.73%
O	6	6	0	C	4.5	CO	10.5	CO	10.5	Volatile	0.375	29.28%	23.56%		
H	3	3	0	C	9	CH_4	12	CH_4	12	Volatile	0.750	58.57%	47.12%		
C	51.4	4.5	46.9	O	6			H_2O (Total)	7.4				25.83%		
		9	37.9	H	3			C	37.9	Non Volatile				37.9	55.41%
Total Kg-moles of VM constituents											1.281	100%			
Total Kg-moles of gases evolved											1.592		100%		
Total Char														68.4	100.0%

*S passes from one solid (coal and/or iron ore) to flux. Hence it is considered non-volatile

Table 5.4 Estimation of iron ore which can be reduced by gases evolved from coal added for generation of only protective char (Basis: 1000 kg or 1 tonne DRI)

Assumptions:

Iron ore analysis: Fe (T) = 66.0 % (all as Fe_2O_3), Gangue=3.7%, Moisture (Free + Combined)=2.0%, Oxygen (with Fe)=28.3%.

Quantity required = 1500 kg

DRI analysis (95% Degree of reduction – Equivalent to 92% Degree of metallisation): Fe(T) = 92.6 %, Fe(M) = 85.7%, FeO = 8.9%, Gangue = 5.2%, C = 0.2%. Quantity = 1069, comprising 1000 kg saleable, 30 kg fine DRI, and 39 kg DRI lost as dust.

Char analysis: Fixed carbon = 55.41%, Ash = 44.59%. 100 kg coal gives 68.400 kg char on carbonisation

Quantity in kilogrammes

Quantity of Iron in DRI	***Quantity of Carbon in Char needed to maintain $C_{fix}/Fe(T)= 0.15$***	***Quantity of Char needed***	***Quantity of Coal needed***	***Gas constituents liberated by this coal***	***Quantity***	***Quantity of Fe_2O_3 reduced by these constituents up to 95%***	***Quantity of Iron Ore reduced***	***Heat deficit, Kcal***	***Product gas***			
									Constituents	***Quantity***	***Kg-moles***	***Percent***
989.89	148.4841	267.977	391.779	N_2	5.093	0	0		N_2	5.093	0.1819	1.40%
				CO_2	1.567	0	0		CO_2	1.567		
				H_2O	28.992	0	0		H_2O	28.992		
				CO	41.137	82.480	87.48	-5,397	CO_2	64.643		
				CH_4	47.013	659.838	699.83	822,422	CO_2	129.287		
				Total			787.31	817,025	H_2O	105.780		
									H_2O from iron ore	15.746		
								CO_2	193.930 (Total)	4.4075	34.0%	
								H_2O	150.518 (Total)	8.3621	64.6%	
									Total gas		12.952	100%

Reduction reactions considered:

$Fe_2O_3+3CO=2Fe+3CO_2$ $\Delta H^\circ_{298} = -11,020$ kCal/kg-mole

&

$4Fe_2O_3+3CH_4=8Fe+3CO_2+6H_2O$ $\Delta H^\circ_{298} = +209,920$ kCal/kg-mole

Table 5.5 Estimation of iron ore which can be reduced by 100 kg of coal

(Constituents converted as per Table 5.3 has been considered here)

Quantity in kilogramms

Sl. No.	Coal constituents	Quantity	Reaction with Fe_2O_3 for complete conversion	Amount of Fe_2O_3 that can be reduced up to 95% reduction	Amount of iron ore being reduced	Product gas			Heat deficit (kCal)
						Constituent	Quantity	Kg-moles	
1	Ash	30.0	-	0	0				
2	H_2O (Total)	7.4	-	0	0	H_2O	45.941	2.5523	
3	CO_2	0.4	-	0	0	CO_2	188.867	4.2924	
4	N_2	1.3	-	0	0	N_2	1.400	0.0500	
5	S	0.5	-	0	0				
6	CO	10.5	(1)	21.053	22.329				–1,377.50
7	CH_4	12	(2)	168.421	178.628				52,480.00
8	C	37.9	(3)	354.620	376.112				116,279.31
			(4)						6,745.29
	Total:	100			577.069			6.8947	174,127.10

(1) $Fe_2O_3+3CO = 2Fe+3CO_2$ $\Delta H^\circ_{298} = -11,020$ kCal/kg-mole

(2) $4F_2O_3+3CH_4 = 8Fe+3CO_2+6H_2O(g)$ $\Delta H^\circ_{298} = +209,920$ kCal/kg-mole

(3) $Fe_2O_3+3/2\ C = 2Fe+3/2\ CO_2$ $\Delta H^\circ_{298} = +55,225$ kCal/kg-mole

(4) $H_2O(l) = H_2O(g)$ $\Delta H^\circ_{298} = +10,520$ kCal/kg-mole

Table 5.6 Estimation of coal heat available from coal for meeting heat requirement (ef. Net Calorific Value)

(Basis: 100 kg coal)

Sl. No.	*Coal constituents (Table 5.3)*	*Quantity*	*Oxygen required for complete combustion (kg)*	*N_2 coming in with the air (kg)*	*Heat Generated*	*Product Gas*		
						Constituents	*Quantity (Kg)*	*Kg-moles*
1.	Ash	30	0	0	0			
2.	H_2O (Total)	7.4	0 (converted to water vapour)	0	–4,324.89	H_2O (g)	34.40	1.9111
3.	CO_2	0.4	0	0	0	CO_2	188.87	4.2924
4.	N_2	1.3	0	0	0	N_2	544.03	19.4298
5.	S	0.5	0	0	0			
6.	CO	10.5	6	21	25,361.25			
7.	CH_4	12	48	168	143,812.50			
8.	C	37.9	101.07	353.73333	297,041.25			
	Total :		155.07	542.73	461,890.11		767.30	25.6333

Reactions considered:

$CO + 1/2\, O_2 = CO_2$ ΔH_{298} = –67,630 kCal/kg-mole

$CH_4 + 2O_2 = CO_2 + 2H_2O(g)$ ΔH_{298} = –191,750 kCal/kg-mole

$C + O_2 = CO_2$ ΔH_{298} = –94,050 kCal/kg-mole

$H_2\, O(l) = H_2O(g)$ ΔH_{298} = +10,520 kCal/kg-mole

Table 5.7 Alternatives considered for calculation along with their boundary conditions

	Ore preheating	*Temperature of preheating*	*Air preheating*	*Mean air temperature*	*Exhaust gas temperature*	*Char recirculation*	*Char quantity recirculated, kg/t DRI*
Alternative-I	No	25°C	No	25°C	1200°C	No	0
Alternative-II	No	25°C	No	25°C	1000°C	No	0
Alternative-III	No	25°C	No	25°C	800°C	No	0
Alternative-IV (Optimised Process*)	Yes	664°C	Yes	400°C	700°C	Yes	200 (~85%)

Conventional process regime lies somewhere between Alternatives 1 and 3.

* *Explained in section 5:13 and Chapter 13*

Table 5.8 Calculation of heat available from the Combustion of 100 kg coal under rotary kiln conditions for the four alternatives

Heat Values in kCal

CV = 4618.9 kCal/kg

Products of combustion of coal			*Heat required to raise the temperature of product constituents*							
Constituents	*Quantity (kg)*	*Kg-moles*	*1200°C*	*Average Cp*	*1000°C*	*Average Cp*	*800°C*	*Average Cp*	*700°C*	*Average Cp*
H_2O (g)	34.40	1.91111	21,214	9.447	17,131	9.194	13,244	8.942	11,374	8.817
CO_2	188.87	4.29242	61,547	12.203	49,870	11.916	38,579	11.597	33,082	11.418
N_2	544.03	19.4298	172,663	7.563	141,341	7.461	110,812	7.359	95,845	7.308
Total	767.30	25.6333	255,424		208,343		162,635		140,301	

Available heat	Alternative1	206,466
	Alternative 2	253,547
	Alternative 3	299,255
	Alternative 4	321,589

Table 5.9 Estimation of heat demands of different alternatives

	Alternative 1		*Alternative 2*		*Alternative 3*		*Alternative 4*	
	Quantity, kg	Heat, kCal	Quantity Kg	Heat, kCal	Quantity, Kg	Heat, kCal	Quantity, Kg	Heat, kCal
Quantity of iron ore reduced by gases evolved from coal added for generation of protective char only, and the heat deficit there from	699.8	822,422	699.8	822,422	699.8	822,422	177.523	208,621
Coal required for reducing the balance iron ore, and the heat deficit there from	138.7	241,447	138.7	241,447	138.7	241,447	229.17	399,050
Calcination of flux	75	32,250	75	32,250	75	32,250	75	32,250
Heat required to raise the temperature of product gases to the outlet temperature		189,530		157,269		125,009		146,789
Radiation and unaccounted losses (assumed)		600,000		600,000		600,000		600,000
Heat put into the system as sensible heat through pre-heated charge (Negative deficit)		0		0		0		–290,998
Total heat deficit		1,885,649		1,853,388		1,821,128		1,095,711
Coal required for meeting heat deficit	913.296		730.983		608.554		340.718	
Coal required for generating protective char	391.78		391.78		391.78		99.38	
Total coal required	**1443.7**		**1261.4**		**1139.0**		**669.3**	

Exit Gas Composition & Quantity

Alternative	*Alternative 2*				*Alternative 4*			
Constituents	H_2O	CO_2	N_2	**Total**	H_2O	CO_2	N_2	**Total**
Kg-moles	25.9	41.73632	142.279	**209.887**	14.482	25.58012	66.361	**106.423**
Percent	12.33%	19.89%	67.79%	**100.00%**	13.61%	24.04%	62.36%	**100.00%**
Gas volume, Nm^3/t DRI	579.5	934.9	3187.1	**4701.5**	324.4	573.0	1486.5	**2383.9**
Air required, Kg-moles/t DRI				177.111				82.55319
Nm^3/t DRI				**3967.29**				**1849.19**

Assumptions (S=Specific heat, Cp=Molar specific heat): $S_{Iron\ ore}$ = 0.225 kCal/kg, S_{Coal} = S_{Char} = 0.38 kCal/kg; S_{Flux} = 0.33 kCal/kg, Heat of Calcination of flux = 430 kCal/kg; Average Cp-product gas = 9.0 kCal/kg-mole

Iron Ore Analysis: Fe (T) = 66% (all as Fe_2O_3), Gangue=3.7%, Moisture (Free+Combined)=2.0%, Oxygen (with Fe)=28.3%. *DRI Analysis* (95% Degree of Reduction – Equivalent to 92% Degree of Metallisation): Fe(T) = 92.6%, Fe(M)=85.7%, FeO=8.9%, Gangue = 5.2%, C = 0.2%. Quantity = 1069, comprising 1000 kg saleable, 30 kg fine DRI, and 39 kg DRI lost as dust. *Coal composition*: Free Moisture = 5.6%, Combined Moisture =1.8%, Ash = 30.0%, C = 51.4%, H = 3.0%, S = 0.5%, N = 1.3%, CO_2 as carbonate = 0.4%, O = 6.0%.

5.9 WASTE GAS TEMPERATURE

To simplify calculations, the present exercise includes that portion of the ABC Chamber in the Sponge Iron Making system where the exit gases have been fully combusted with stoichiometric amount of air. The portion of ABC, where the gas is further diluted with excess air and is then cooled by water spray, is excluded from calculation.

Since the two functions within the ABC chamber are overlapping it is extremely difficult to estimate, let alone measure, the temperature of the gas after complete combustion and before it is diluted with excess air and cooled with water spray. A reasonable guess is that this temperature lies between 800°C to 1200°C. Comparison of results of modelling with those of actual operating plants indicate that modal value of this temperature is about 1000°C, i.e., a majority of the plants operation is consistent with the waste gas temperature of about 1000°C after full combustion.

5.10 PROGRAMMING BASED ON MODEL

Up to the above we have conceived of a framework of calculation of the overall process parameters. This framework of calculation is often represented in a 'flowchart' and has to be converted to a 'program' for a computer to be able to calculate out results from certain inputs. Earlier with programming language like "Basic" or "Fortran" it was essential first to construct the 'flowchart'. The 'flowchart' was (and is) referred to as the 'model' as separated from the 'program' which followed.

With the advent of object oriented programming (using computer languages like 'C', 'C++' etc.) making flow chart is no longer an essential part of programming. But making a flow chart is advisable to easily understand a model.

5.11 SPREAD-SHEET PROGRAMMING

Tables 5.3 to 5.9 together constitute an overall model of a typical sponge iron making in rotary kiln and the programming and solutions are based on the 'spread sheet' method ('Lotus' 1-2-3 or, as in this case, 'Microsoft Excel'). The spread-sheet method is not the very best way of writing computer program of this type, but can be an easy and convenient step towards understanding a model. Tables 5.3 to 5.9 combine the material and energy balance of the process with typical inputs. Many of the desired outputs can be projected or estimated, such as energy consumption, energy in waste gas, waste gas composition, air requirement, quantity of coal required, etc. Please note that Tables 5.3 to 5.9 are based on an iron ore composition different from that used in Tables 5.1 and 5.2.

5.12 SEGMENT-WISE MODELLING

Having developed an overall model, we would like to extend it so that we can estimate what happens at the intermediate stages, such as at the middle, at one third or two third length etc. We would like to know along the length of the kiln, the reduction or metallisation profile, the temperature profiles of the bed and gas, the residence time required, etc. For this purpose we resort to "segment-wise modelling" illustraed in Chapter 7. We divide the kiln into a large number of segments, and we calculate the parameters in each segment and the result becomes an input to the next segment. Within the segment we assume uniform conditions of temperature of gas, temperature of solid bed, metallization, gas flow, etc. For this assumption to be reasonably valid, we have to reduce segment size, in successive calculations, and select such a size of segment where size change does not significantly affect the results of calculation.

The kinetics of reduction would be an important input to this exercise. The kinetic parameters like the iron ore reducibility and coal char reactivity are widely different for different ores and coals. These properties keep changing with exposure to different temperatures and for different lengths of time in the rotary kiln. It is very difficult to accurately estimate the reaction rates within the rotary kiln segments using the basic kinetic properties of the inputs, and this approach very often does not lead to the development of a practical mathematical model.

A practical mathematical model would instead use the metallisation profile generated in laboratory rotary tube furnace using the particular iron ore and coal combination.

5.13 PREDICTION FROM THE MODEL

The results from a typical mathematical model calculation based on spread sheet programming are presented in Tables 5.3 to 5.9. These results and its extensions into segment wise modelling have indicated a few important things:

1. When a rotary kiln is operated with 100% thrown coal, there is about 40% saving in coal consumption, as compared to the normal practice of using 40% thrown coal and 60% feed coal. However, there would be some decline in productivity.
   This result was validated in Pilot Plant experiments as consumption was found to decrease from 938 kg/tonne sponge iron to 562 kg/tonne sponge iron.
2. Extending the above finding, the model has further predicted that if a preheating kiln is incorporated

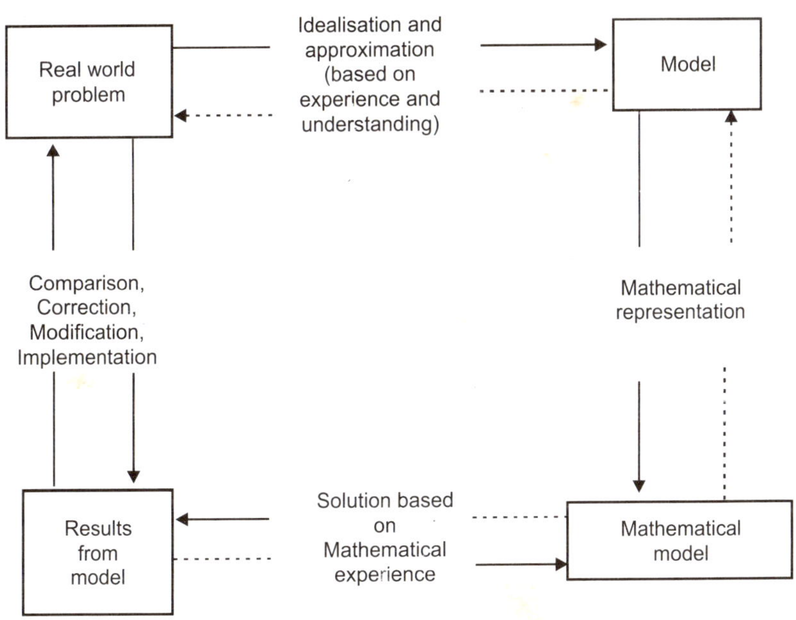

Fig. 5.1 Schematic representation of the process of mathematical modelling

upstream of the reduction kiln, where char is also re-circulated, and the resulting waste gas from the preheating kiln is used to pre-heat air, which is ploughed back into the process, the coal consumption would reduce by 50%. This concept has been referred to as the optimised process in Table 5.7 and has been discussed in detail in Chapter 13.

3. For a more complete exercise, if we want to estimate the rate of revolution and the kiln inclination, we need further to develop the charge movement and residence time models. To accurately estimate the gas temperature profile required and consequently the air requirements in different segments, we need to develop the heat transfer model. Both these exercises are also quite involved and are dealt in the subsequent sections.

5.14 SUMMARY OF THE MODELLING PROCESS

Figure 5.1 adequately depicts the different interconnected steps involved in the development of a practical, workable and useful mathematical model of a process such as the process of sponge iron making in a rotary kiln. Such a mathematical process, once developed, helps in understanding of the process, process simulation and predicting various outputs for different inputs. These data can be generated without doing costly experiments. But the reliability of these results depends heavily on how well the model has been validated.

Further, any model is as good as the assumption made. There is need to keep a continuous check on whether the assumptions, which were valid during model validation, have remained valid during application or not.

In the present case of mathematical modelling of the coal based rotary kiln sponge iron making process, the model developed and its application has helped, among other things, in better understanding of the process. This has also pointed towards some design limitations of the equipment presently being used. From these results, it has been possible to develop an optimized process concept with a preheating kiln, an air pre-heater and use of char within the process itself.

Physical Movement of Solids Inside a Rotary Kiln: Charge Movement and Coal Throwing/Slinging

6

CHAPTER

Sponge iron making in rotary kiln involves reduction of granular iron ore in solid state by the reducing gases generated by the gasification of coal. After the upper layer of iron ore is reduced, the reduction of the interior requires diffusion of the reducing gases, mainly CO, to the reduction front and diffusion of the CO_2 formed, out to the surface and freeboard (Fig. 1.7 and 2.9).

For the process to be commercially viable the solids need to be exposed to the reducing gases at as high a temperature as can be tolerated for as long a period as possible. Since fusion of constituents causes multiple problems in the kiln, the maximum temperature of solids is controlled to an optimum level, generally around 1000°C (Fig. 2.1)

6.1 IMPORTANCE OF RESIDENCE TIME

For designing such a sponge iron kiln, one needs to know the residence time requirement and the correlation of the residence time with the inclination, speed of rotation (revolutions per minute, i.e., rpm) and the dimensions of the kiln. While most of the early kilns decided on their inclination and rpm based on their experience in getting good quality product, newer kilns are based on much more scientific design data. Residence time requirement is one of the more important design data for such kilns.

Residence time refers to the time spent by the solid granular charge in travelling from the charging end to the discharge end of the rotary kiln. There is some weight loss and, to a lesser extent volume decrease. In our treatment, volume considerations are more important, and bulk density based on mean volume has been used in further treatment.

One expert had remarked that it is very easy to conceive that if we put some granular material at the higher end of a hollow inclined cylinder, and we rotate it, the material would eventually come out from the lower end. But if we try to examine the mechanism, we find it highly complicated.

6.2 ESTIMATING GROSS RESIDENCE TIME FROM INPUT AND OUTPUT OF SOLIDS

In a sponge iron rotary kiln the material is likely to follow a path represented in Fig. 6.1. About 15% average filling degree of the kiln is generally maintained in the sponge iron rotary kilns. Approximately six hours reduction time is required for one of the better combinations of ore and coal. The bulk density of the charge mix generally lies between 1.6 to 1.8 t/m^3 while that for discharge material, mostly sponge iron and char, it is between 1.3 and 1.5 t/m^3. We may use an average figure of 1.4 t/m^3 as an example. About a tonne of sponge iron in discharge is accompanied with about 400 kg of char and ash, making the total discharge per tonne sponge iron as 1.4 tonnes or 1.0 m^3.

Let us take the example of a 50-tpd sponge iron kiln. It would have to discharge about 2 m^3 per hour of material or about 16 m^3 in eight hours. That means, if a total of eight hours residence time is envisaged (two hours for preheating), the solid material at any instant would occupy 16 m^3 volume in the kiln. Assuming average filling degree as 15% the total inner volume of kiln after lining should be about 110 m^3, which is a very important design data of the kiln.

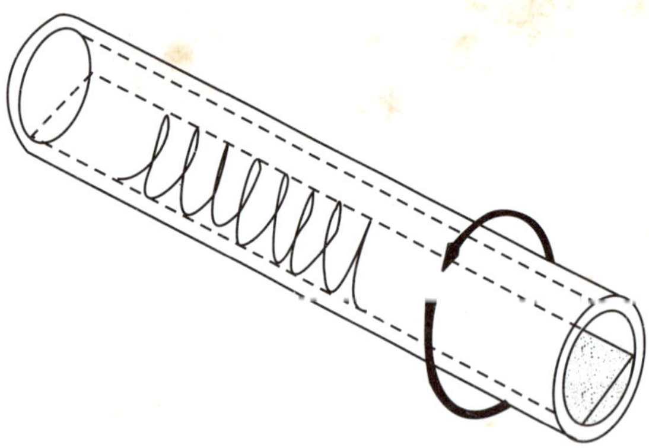

Fig. 6.1 Material movement in a rotary kiln

Here we have assumed the residence time of material in the kiln. In actual practice, the plant operators try to drive the kiln as hard as possible while maintaining the quality of the product and simultaneously looking after the safety of the kiln. By using the above calculation method for an operating kiln, one can estimate the residence time while making assumptions on the average filling degree and the average bulk density. To check whether these assumptions are reasonably correct one needs to determine the residence time precisely.

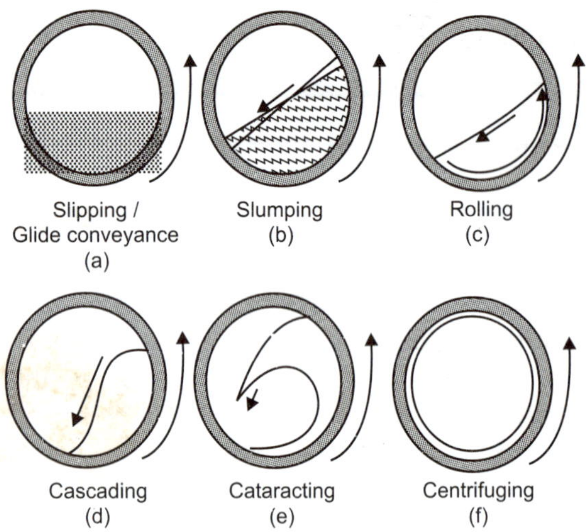

Fig. 6.2 Modes of transverse bed motion in a rotating cylinder

6.3 CHARGE MOVEMENT: CASCADING FLOW OF GRANULAR MATERIAL

If we keep some granular material in a horizontal hollow rotary cylinder and we rotate the cylinder axially, the material would behave differently for different conditions of surface friction and speed of rotation.

By the term granular material we mean a solid material which is neither too fine in size (i.e. it is not finely ground) nor is it in as mined (boulder) form. Rather it is in a close size range such as 10 to 40 mm, 8 to 30 mm, 6 to 20 mm or 5 to 18 mm – the sizes used in a blast furnace or a sponge iron rotary kiln.

If, in the above example, the cylinder is slightly inclined, we get conditions similar to that existing in a rotary kiln or a rotary cooler. The six different behaviours of the granular material are represented in Fig. 6.1 – (a) to (f).

Condition of Fig. 6.1(a) occurs when the friction between the charge material and the cylinder surface is very low. The charge material slips on the cylinder surface (Slipping); the rotation of the cylinder does not have any effect on the material. If the cylinder is axially inclined the material just glides down the inclination (Gliding Conveyance). Readers may consider this situation as not of practical importance – and indeed this would not happen on a refractory surface as that would be sufficiently rough – but this does happen when refractory is not present, as in a rotary cooler. Initially, rotary coolers, which were meant to cool sponge iron indirectly, out of contact with both air and water, consisted of just a cylindrical shell of steel, sprayed from outside with water. With continuous abrasion due to movement of sponge iron, the inner surface used to acquire a mirror finish, leading to gliding conveyance and thereby insufficient cooling of sponge iron. Often the upper surface continued to remain red hot at discharge. In subsequent designs, lifters became integral part of cooler design which eliminated this problem.

Figure 6.1(b) represents a situation where the roughness of the cylinder surface carries the material up with its rotation until a point is reached where the static angle of repose is exceeded. Angle of repose is the highest angle at which a granular material can rest in a stockpile. We have used the term static angle of repose to distinguish it from the dynamic angle of repose which is the angle of repose in a slightly disturbed condition and obviously has a value slightly lower than the static angle of repose. When the static angle of repose is exceeded, a small quantity of material slumps from the upper side down to the lower side to roughly make the dynamic angle of repose. After settling in this position the material remains stationary with respect to the cylinder surface until the static angle of repose is reached once again.

The above happens when the speed of rotation of the cylinder is slow enough to give the material a resting time between slumps. If the rotation is fast enough so that the static angle of repose of material is reached before the slumping is completed, then there is no time for the material to rest. We get a movement of charge depicted in Fig. 6.1(c), which most researchers like to call rolling movement. Some others have termed it trickling movement. As we would see later, this type of movement is the most desirable in a sponge iron rotary kiln.

As the rotational speed of cylinder is progressively increased further, material movement changes to cascading (Fig. 6.2 (d) – some additional voids created within the bed) to cataracting (Fig. 6.2 (e) – also called detaching or slinging) to finally centrifuging (Fig. 6.2 (f)). In the last named condition, centrifugal force makes the material to stick to the cylinder surface and any relative movement between the material and the inner surface of cylinder completely stops

In the first and the last condition, the mixing action of material, which is one of the main strengths of rotary kiln, completely stops and therefore are not desirable in a rotary kiln. In the condition (b) i.e., cascading, the mixing action is not continuous, as also the surface renewal. The latter would somewhat impair the heat transfer to the upper surface of solid material. Henein *(1983)* determined conditions which decide the change in mode of movement from cascading to rolling. This happens at different rpm's for different diameters of kiln and the nature of material, but for commercial rotary kiln sponge iron plants the rpm value hovers around 0.3. For small diameter kilns the value is slightly larger while for large diameter kilns the value is very marginally lower. So, 0.3 is about the very lowest value of rpm at which the rolling action is obtained.

Cascading and cataracting actions do not give any special advantage over rolling action in the context of sponge iron making. On the contrary, overlapping between solid and freeboard starts which is undesirable from reduction point of view. Also higher rpm increases electrical power consumption as also wear and tear on the machinery and is, therefore, avoided. Some flexibility in rpm variation is needed to increase or decrease residence time and, therefore, commercial kilns are designed to operate between 0.3 to 0.6 rpm and generally between 0.4 to 0.5 rpm.

6.4 MATHEMATICAL TREATMENT OF CHARGE MOVEMENT

Let us assume that a particle at point A (Figure 6.3) moves up to the point B with rotation of kiln and then slides down. Due to the action of gravity, the particle, instead of sliding down to point A, comes down to a point K as shown in Fig. 6.3(a). Let us call this cycle of bed rotation as one "Cascade" (not to be confused with cascading action mentioned in the previous section). Therefore, due to one sliding movement, or in one Cascade, the net forward movement of the particle is AK.

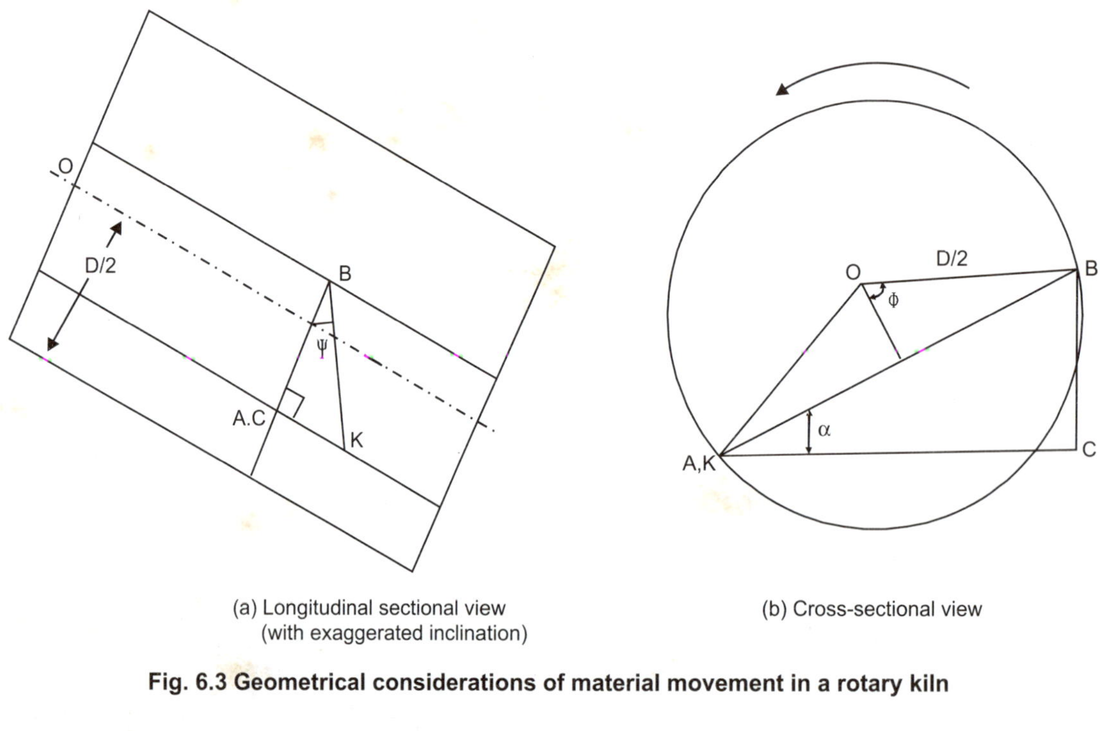

(a) Longitudinal sectional view (with exaggerated inclination)

(b) Cross-sectional view

Fig. 6.3 Geometrical considerations of material movement in a rotary kiln

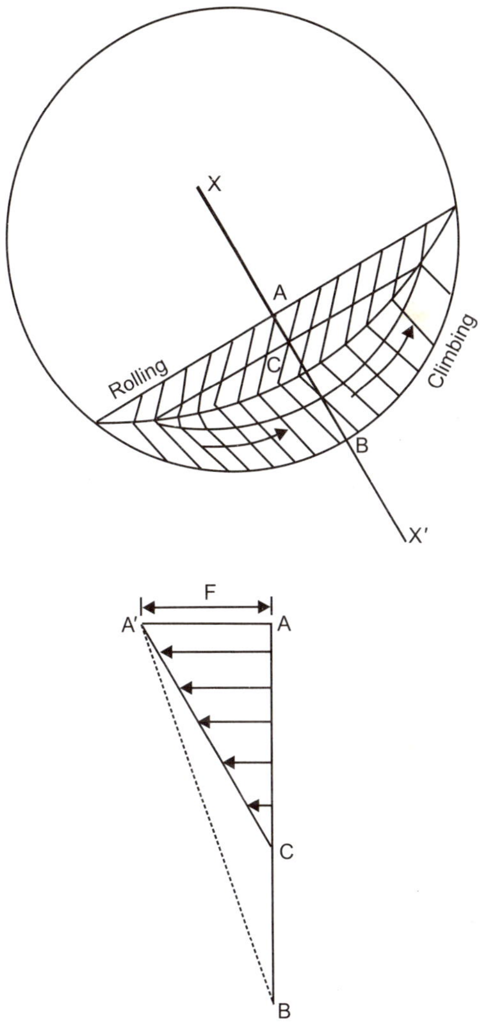

Profile along X-X′

Fig. 6.4 Particle movement profile in a charge bed due to climbing and rolling

From Fig. 6.3(a),

Forward movement (AK)

$$= BC \tan \psi \qquad (1)$$

where ψ = angle of inclination of kiln with horizontal plane.

As shown in Fig. 6.3(b), the angle α is the static angle of repose of the bed material and ϕ is the semi-vertical angle subtended by the charge bed at the centre of the kiln. α should, in fact, be greater by an amount equal to $2\pi^2DN^2\sin\phi/(\phi^2\ g\cos\alpha)$ than the angle of repose, but this is negligible compared to α. Further, in the rotary kiln condition the dynamic angle of repose is more relevant and this value is slightly less than α. Both these differences tend to annul each other and in any case they are small and can be neglected. We are, therefore, reasonably justified in using the static angle of repose in this formulation.

From Fig. 6.3(b),

$$BC = AB \sin\alpha$$

$$= D \sin\phi \sin\alpha \qquad (2)$$

And forward movement

$$AK = BC \tan\psi$$

$$= D \sin\alpha \sin\phi \tan\psi \qquad (2(a))$$

Let N be the speed of revolution of the kiln in revolutions per minute (rpm). Hence, time spent per cascade

$$= 2\phi/(2\pi N)$$

$$= \phi/(\pi N) \quad \text{(minutes)} \qquad (3)$$

Forward movement of the particle per minute

$$= (D \sin\phi \sin\alpha \tan\psi)/(\phi/\pi N)$$

$$= (\pi DN \sin\phi \sin\alpha \tan\psi)/\phi$$

Let the length of the kiln be L.

Therefore, time required in covering distance L,

$$T = L/((\pi DN \sin\phi \sin\alpha \tan\psi)/\phi)$$

$$= L\phi/(\pi DN \sin\phi \sin\alpha \tan\psi)$$

Thus the residence time (T) should apparently be

$$= (L\phi/(\pi N))/(1/(D \sin\phi \sin\alpha \tan\psi)) \qquad (4)$$

6.4.1 Testing of the Mathematical Expression and its Amendment

When attempts were made to calculate residence time in the rotary kiln, both commercial and experimental (cold and hot), the values obtained were much higher than the realistic estimates based on filling degree and bulk density.

Further, the above formulation is based on the movement of particles on the surface layer. The forward movement of particles in the sub-surface layers should be proportionately lesser and the average forward movement should be reduced by a factor of half (Fig. 6.4), which would predict a residence time which is much more than double of the realistic estimates.

It was reasoned that the explanation of this difference in estimates lies probably in the movement caused by bed disturbance. A correction is therefore required to be applied for this factor.

6.4.2 Bed Disturbance Factor

If we consider a heap of granular material (Fig. 6.5(a)) and disturb it by, say, vibrating the base, the material would flow on both sides, if the base is horizontal (Fig. 6.5(b)). If

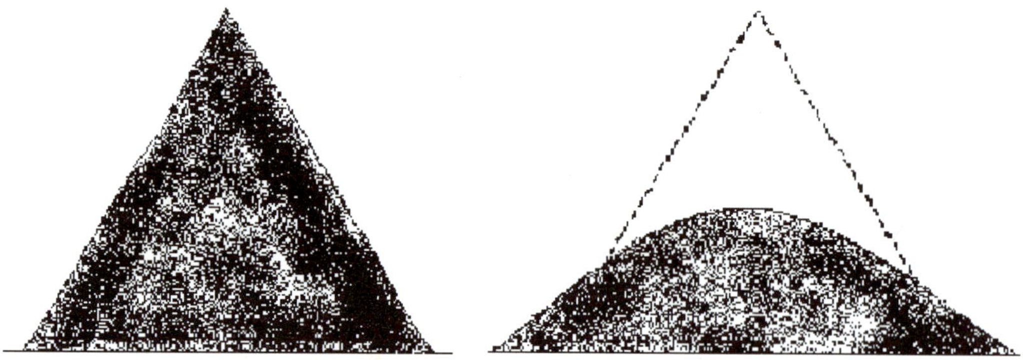

(a) Material heap before being disturbed (b) Material heap after being disturbed

Fig. 6.5 Flow of material caused by disturbance

Table 6.1 Formulations for residence time in a rotary vessel

Sl. No.	*Author/ organisation*	*Formulation for residence time, minutes*	*Type of formulation*	*Remarks*
1.	Sullivan et al., U.S. Bureau of Mines	$5.81\ L\alpha^{0.5}\ C/(DN\ \psi)$	Theoretical + Empirical	C is a factor representing constictions. Generally equal to 2
2.	Ginstling et al	$0.4698\ L/(DN\ \psi)$	Theoretical + Empirical	
3.	Saeman	$L \sin\alpha/\pi\ DN(\psi+\gamma\cos\alpha)$	Theoretical	α, ψ & γ are in radians.
4.	Vahl and Knigma	$0.91\ (L/D)^2\ (D/h_0)^{0.57}\ (\tan\alpha\ /N))$	Theoretical + Empirical	For horizontal hollow cylinders
5.	Chaterjee et al	$k(L^3/F)(\alpha/\psi)^{1.054}(F/NL^3)^{0.981}$	Statistical	Based on Dimensional Analysis
6.	Bandopadhyay et al	$L\phi/(\pi DN\ (\sin\alpha\sin\phi\ \tan\psi + K/N))$	Theoretical + Empirical	

the base were inclined, the material would flow more towards the inclination. Thus, material flow can occur in an inclined rotary kiln even without the geometrical effect of rotation, and just by providing disturbance to bed. In a sponge iron rotary kiln, the cascading movement of the material always disturbs the bed.

Thus, it can be concluded that the movement of charge material from the upper to the lower end of the rotary kiln is due to the following two factors.

(i) Forward movement which can be calculated by the geometry of cascading; and

(ii) Additional forward movement due to constant disturbance of the bed. Let us denote this additional movement as f_p.

Thus, we can state that,

Forward movement per cascade

$$= D \sin \phi \sin \alpha \tan \psi + f_p$$

f_p is expected to depend on the various operating parameters like ϕ, N, D, ψ, etc. The formulation of f_p may take a number of forms depending upon the logic that can be applied for explaining the additional forward movement. Since these possible logics are little better than conjectures, it was decided to try out a large number of formulations and then to observe as to which formulation explains the experimental results better.

The experiments towards evaluation of the above were carried out in cold miniature models of rotary kiln. A number of formulations were tried out for f_p and that particular formulation was selected which could predict the results of the experimental observation with the minimum variance of error. Based on this analysis the best formulation of f_p was found to be

$$f_p = KD/N$$

where K is an empirical constant.

Hence,

Forward movement per cascade

$$= D (\sin \phi \sin \alpha \tan \psi + K/N)$$

and

Residence time, T

$$= L\phi/(\pi DN (\sin \phi \sin \alpha \tan \psi + K/N))$$

This formulation is a combination of theoretical and statistical analyses. One may like to compare this formulation with others available in literature (Table 6.1).

This formulation was used to estimate the residence time of material in a hot pilot sponge iron kiln of nominal dimension of 1.3 m diameter and 12 m length. To know the actual residence time a few specialised techniques were needed which are presented in subsequent section. The measured values and those, which can be calculated using the above model, are compared in Table 6.2.

6.5 IMPORTANCE OF FILLING DEGREE

For commercial production purpose, operators would like to load the kiln with as much solids as possible. But, as mentioned earlier, sponge iron making in a rotary kiln is a delicate balancing exercise, particularly in the zone where FeO is being reduced to metallic iron. This has been illustrated in Fig. 2.9. Highly reducing condition is required within the solid bed so that the CO/CO_2 ratio never falls below the limiting value of 2.5 and the CO_2 formed as a result of reduction is quickly reconverted to CO. On the other hand, in the gas phase, substantial quantities of tri-atomic gases (gas molecules containing three atoms, e.g., CO_2, H_2O) are needed so that heat transfer by radiation from gas is not

Table 6.2 Comparison of the experimentally determined residence time (during iron ore reduction) with the predictions from the model (i.e., formulation no. 6 of Table 6.1)

Expt. No.	Tracers used		Residence Time, Minutes			Problems faced, possible reasons for deviation and remedial actions taken
	Physical or Chemical	Details	Measured	Predicted by Model	Difference (Predicted-Measured)	
1.	Physical	Pellets	343	284	– 59	Feeb impulse.
2.	Physical	Fine iron ore	328	297	– 31	Impulse too small to be easily discemed . Quantity of iron ore increased to 30% kg.
3.	Physical	Fine iron ore	293	294	+ 1	
4.	Physical	Fine iron ore	297	307	+ 10	
5.	Physical	Fine iron ore	311	309	– 2	
6.	Chemical	Manganese ore	314	279	– 35	Impulse too small and peak identity could have been mistaken, Manganese ore quantity increased to 60 kg.
7.	Chemical	Manganese ore	278	291	+ 13	
8.	Chemical	Permanganate impregnated ore	322	315	– 7	
9.	Chemical	Calcium sulphate impregnated ore	264	284	+ 20	Feeble impulse. Use of calcium sulphate discontinued.
10.	Chemical	Copper sulphate impregnated ore	243	234	– 9	

hampered. Diatomic gases like CO, H_2, O_2 and N_2 do not radiate heat (or rather the heat radiation is negligible). If the CO_2 and H_2O molecules come near the solid bed, they would not only stop the reduction process but also would re-oxidise the already formed metallic iron.

In order to maintain the presence of substantial amounts of CO_2 and H_2O molecules in the gas phase without disturbing the reduction process, a large free board (a place where there is no solid) is maintained and the process air is introduced at a location which is far removed from the solid bed. It has been found in practice that to maintain an effective rate of reduction, solids have to be filled to a level which is not more than 15% of the cross-section. This is one major reason why the productivity of sponge iron rotary kilns is much much lower than that of, say, blast furnace. But this does not make rotary kiln unviable since it does perform multiple functions not possible in many other metallurgical reactors.

6.6 TECHNIQUES OF MEASURING RESIDENCE TIME

A number of techniques have been employed in the past to measure the residence time in actual sponge iron rotary kilns but unfortunately, these have not been documented. Through informal sources, we have come to know of the following techniques of measurements:

(1) Use of 'Markers'

Inert materials of large sizes have been charged and attempt has been made to trace their passage through the hot kiln. It is virtually impossible to see inside an operating kiln without stopping it, as heavily dusty atmosphere, caused by flying tiny ash particles, exists inside all the time. Markers can be tracked only when they reach the discharge end. As there is always a dam at the discharge end in a sponge iron rotary kiln, the marker, which is larger than the normal sized material, finds it difficult to cross over the dam. The results have mostly given inflated values and only lower of a series of results can be relied upon. However, this cannot be considered a scientific method of determination of residence time.

(2) Monitoring of Time Gap Between First Charging and First Discharging:

This time gap also gives only a rough idea of residence time as the filling degree over the length has not stabilised and this data does not reflect the situation under steady state condition.

Only these two techniques have been employed to our knowledge in actual sponge iron kilns. More often than not, agencies transferring the technology have provided statistical formulations to their clients and have asked them to calculate the residence time accordingly. The need for knowing the actual residence time was not felt necessary while operating a plant. But this value is required when one goes to design a new kiln.

6.6.1 Measurement of Residence Time in the Pilot Rotary Kiln of RDCIS SAIL

The dimension of the kiln, as mentioned in Chapter 2, was 1.3 m dia and 12 m length. One effective way of measuring the residence time here could have been by using radioactive tracers. But this involves a safety hazard and the idea of using it was dropped. It was decided instead to use physical and chemical tracers in our experiments.

(a) Physical Tracers

Pitfalls of using a physical tracer like the 'marker' described above were realised in the beginning. But a different physical tracer such as iron ore fines is expected to give a value better than that by marker technique. To get rough idea of the residence time, iron ore fines were used as the physical tracer in one set of experiments and small sized iron ore pellets were used in another set. Two graphs of such determinations, one for the reduction kiln together with the rotary cooler and the second only for the rotary cooler part of the Pilot Plant, have been presented in Fig. 6.6 & 6.7. While experiments with iron ore fines yielded significant impulses in the size plot of discharge material, pellets gave very poor impulse possibly due to easy flowability leading to pronounced back mixing.

(b) Chemical Tracers

It was realised early that for more accurate measurement of residence time one must use chemical tracers keeping the physical properties as far unchanged as possible, as the presence of even extra fines, for example, can change the movement pattern. Towards this end, the following were attempted as chemical tracers.

(i) Sulphur as chemical tracer

Iron ore lumps were soaked in calcium sulphate solution and dried. They were then charged for a specified time instead of the normal iron ore. The product was regularly

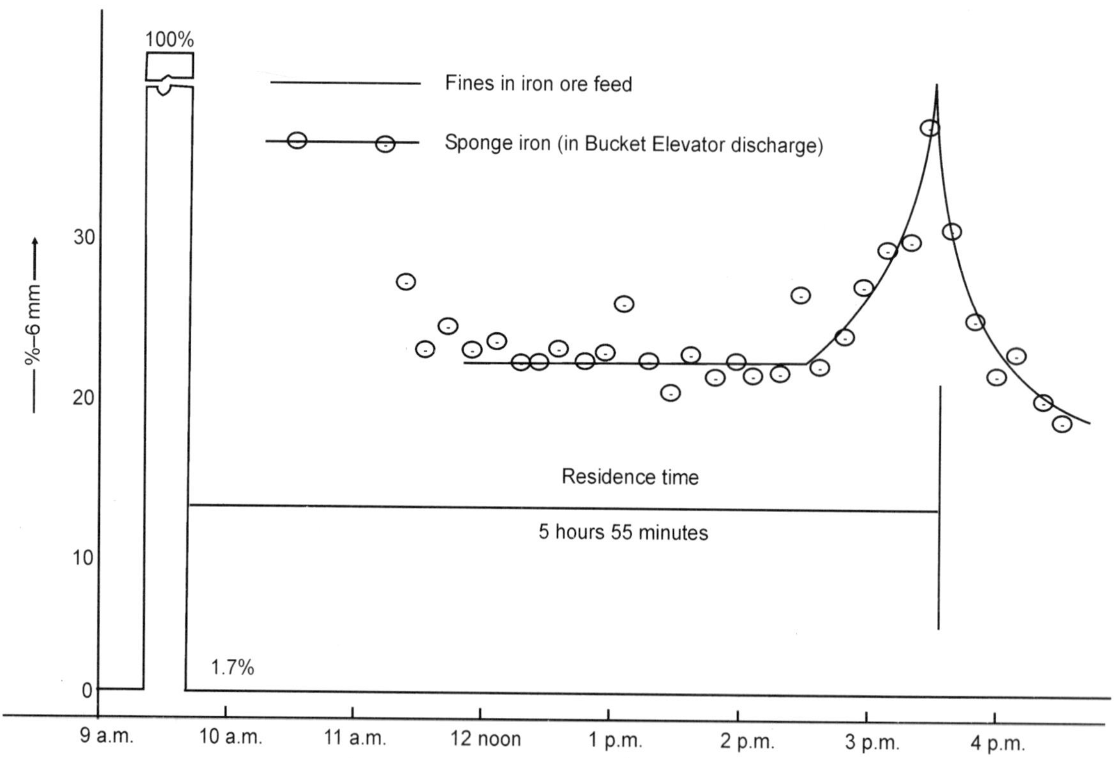

Fig. 6.6 Residence time plot for an experiment with fine iron ore as tracer covering both rotary kiln and cooler

analysed for sulphur content in a Leco C-S analyser. The time gap between the mean sulphur impulses from the charging point to the sampling point gave the time of travel between these two points. In this case, however, the impulse obtained was feeble.

(ii) Carbon as chemical tracer

Cold bonded pellets impregnated with coal were prepared of roughly the same size as iron ore. Iron ore was replaced with these pellets at a given time and the magnetic product was analysed for carbon content, since the composite pellets were expected to give sponge iron containing higher carbon. Residence time was determined as before.

(iii) Manganese as tracer

Manganese was used as a chemical tracer in two ways. In the first set of experiments, iron ore was soaked in a strong permanganate solution, dried and then used. In another set of experiments, Manganese ore itself was crushed to the same size as iron ore and then charged in the kiln.

The composite product was analysed for manganese and the concentrations plotted. Figure 6.8 gives the plot of one particular experiment with Manganese ore as tracer.

(iv) Use of copper and sulphur simultaneously as chemical tracers

Iron ore was soaked in strong copper

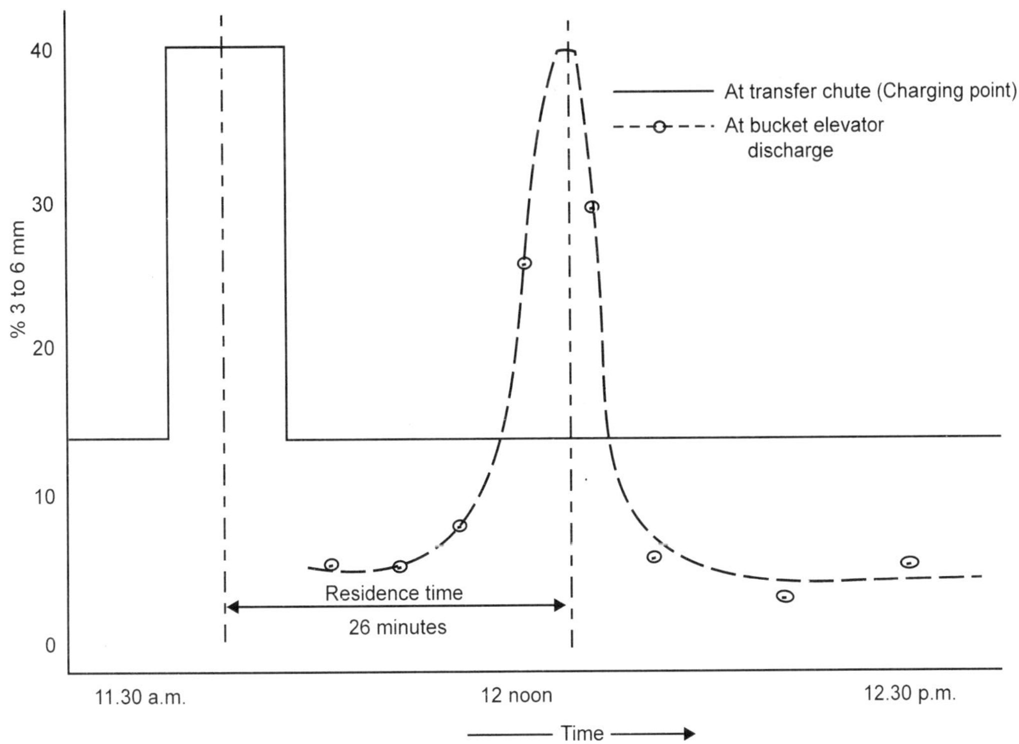

Fig. 6.7 Residence time plot for an experiment with fine iron ore as tracer in the rotary cooler

sulphate solution and dried for the purpose of residence time measurement. Surprisingly the copper concentration impulse in the product was not significant. But this technique proved to be a more successful variation of the sulphur tracer technique.

The success of all the experiments depended on getting a measurable impulse of the tracer at sampling point. In some of the early experiment, no tracer impulse could be detected even after careful sampling and analysis at close intervals. The effect depicted in Fig. 6.5 also caused back mixing of the different material charged with the regular material of the kiln, making the impulse flatter and flatter with travelling of larger distances and passage of time.

For the Pilot Plant condition, and while using iron ore as feed material, pellets were found to be a poor tracer, as even 30 kg of input could not given a significant impulse in the product size plot. However, pellets were found useful as tracer when pellets (of different kind) were used as raw material. These results are not presented here.

Iron ore fines as tracer could give significant impulse in the product size plots even at 30 kg or six minutes of feeding. To have a measurable impulse with chemical tracers, 12 to 15 minutes of feeding of the tracer containing inputs was found to be necessary. But with iron ore soaked in calcium sulphate, even higher quantity of feeding was requiıed for obtaining a measurable sulphur impulse. For a commercial kiln, an even higher tracer input impulse may be required for a detectable and reliable output impulse.

This technique is simple in concept, and the results can be utilised for optimising operations of an existing kiln.

6.7 IMPORTANCE OF THROWING COAL FROM DISCHARGE END

The final stages of reduction in a sponge iron

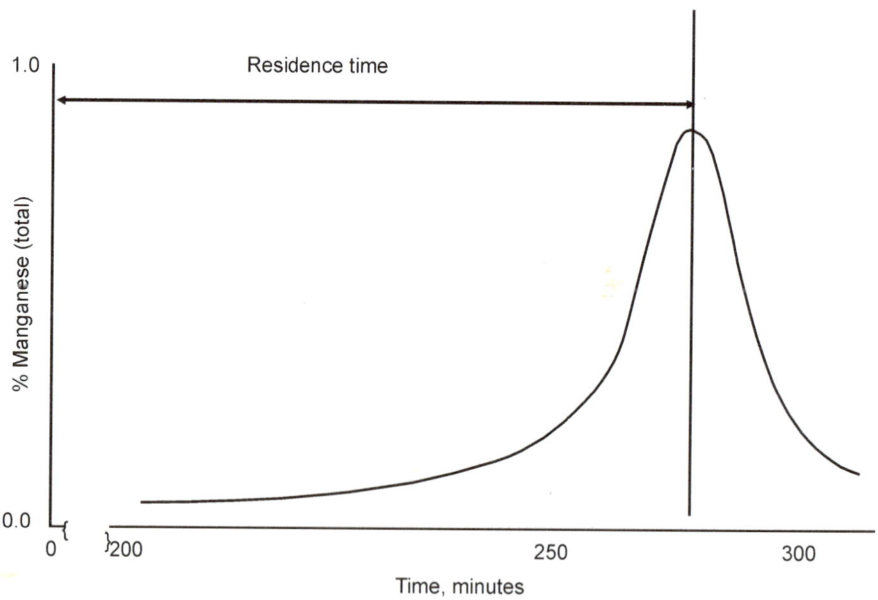

Fig. 6.8 Residence time plot obtained while using manganese ore as tracer

rotary kiln are the most important as well as the most difficult. As FeO concentration reduces, the rate of reduction drops. Further, since the char in the solid bed, formed from the feed coal, has spent a few hours at high temperature, has lost much of its reactivity, i.e., its capacity to convert CO_2 quickly back to CO. This also reduces and, beyond a limit, stops the reduction process.

If fresh coal is introduced during the last stages of reduction, fresh reactive char can be made available, which can enhance the reduction rate and thus greatly enhance the viability of the process. This is done by propelling some coal (throwing or slinging) to some distance in the reduction zone.

Since the solid charge is continuously moving forward, dropping coal at the kiln mouth would be of little use since coal would immediately accompany the kiln discharge into the cooler. Coal has to be projected or propelled to some distance in the reduction zone.

Throwing coal from discharge end has now become standard practice in almost all variations of rotary kiln sponge iron making. Air is the universally adopted medium for carrying and giving momentum to coal particles, as air is anyway needed within the kiln at this location. Since too much air would render the atmosphere too oxidising to disturb the reduction condition within the bed, throwing has to be achieved with controlled amount of air.

From the weigh feeder (or the volumetric feeder as the case may be) coal would fall in irregular spurts. If this is led directly to the pipe carrying air, there is every chance of blocking the air pipe. To avoid this, coal feed has to be regulated through a feeder, which would release coal in small quantities at regular intervals. This is done by introducing a star feeder (having grooves arranged around a shaft in a star fashion) or some variation thereof.

Air in the coal throw pipe comes with a certain velocity governed by the flow rate and the cross-sectional area of the pipe. Pipe is kept roughly parallel to the axis of the kiln. Coal particles initially have zero velocity in

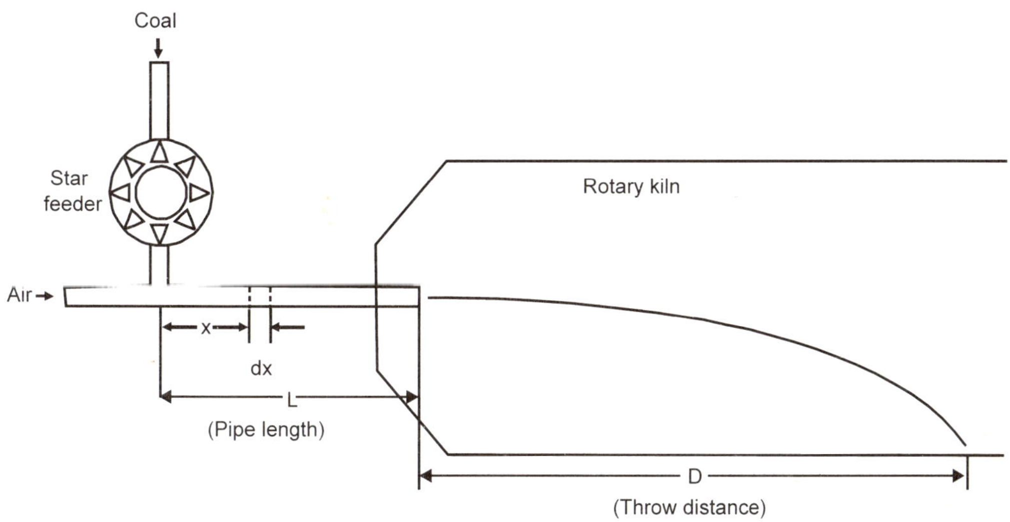

Fig. 6.9 A typical coal throwing arrangement

the direction of the kiln axis. The air flow drags the coal particles along by its velocity. The coal particles gain acceleration within the throw pipe depending upon the particle surface area, its weight and the difference in the velocities between the particle and air. The difference in velocity keeps decreasing as the particle gains velocity and therefore acceleration rate keeps decreasing, until the time when the mixture comes out of the pipe. Air velocity drops to negligible value and the velocity of the projected coal starts decreasing, until it falls by gravity on the solid charge bed.

(a) Throw distances with 65 mm air pipe & 7 metre pipe length

(b) Throw distances with 65 mm air pipe & 10 metre pipe length

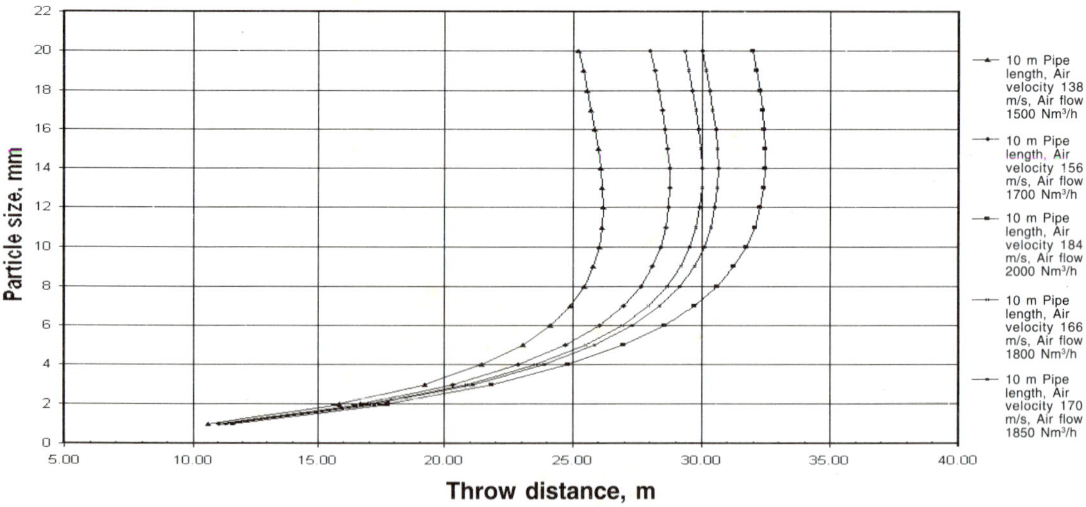

(c) Throw distances with 65 mm air pipe & 15 metre pipe length

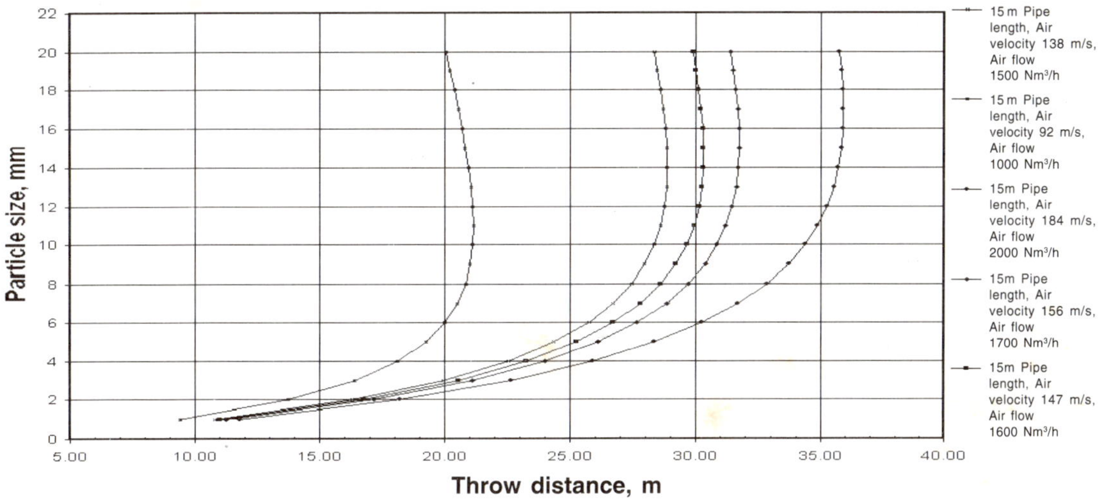

Fig. 6.10 Calculated distances of throw of coal particles under different conditions

6.8 COAL THROWING PHILOSOPHY

Air velocities used in a coal throw pipe are quite high and always in the turbulent zone. Turbulent flow formulations would apply. As per this,

$$F_D = 0.44^* \pi ^* d_p^{2*} \rho^* (V_a - u)/8$$

where,

F_D = Drag force on the coal particle

d_p = Coal particle diameter (Assumed spherical)

ρ_a = Density of air

V_a = Velocity of air

u = Velocity attained by coal particles at any distance x within the pipe

Acceleration rate of a coal particle is given by

$$fc = F_D/m$$

where,

m = Particle mass

On integration between limits when time t = 0 also u = 0 and till any time t, we get, for a particular case,

$$1/(V_a - u) = 1/V_a + 0.02394^*t$$

Here we have assumed the following values as an example,

d_p = 12 mm

V_a = 120 m/s

ρ_a = 1.21875 kg/m^3

ρ_p = 1400 kg/m^3

From these results it can be calculated that for the coal particle (12 mm spherical) to attain 25% of the air velocity only 1.9 m pipe length is enough. If 50% of air velocity is to be attained, 12.8 m pipe length is required. For attaining 75% or higher velocity the length requirement becomes prohibitively high, 67.4 metres or more.

Once the coal particle emerges out of the throw pipe, the forward drag force stops.

Instead, the particle is dragged back by the practically stagnant gas in the kiln freeboard. Similar considerations in the reverse direction apply. If, in this particular case, the free fall vertical distance available to coal particle is 1.6 m, the calculated distances covered by the particles of various sizes are presented in Fig. 6.10. It can be observed that air velocities can be substantially reduced for the same distance of throw if longer pipe lengths are used. This fact has been applied in the sponge iron rotary kilns based on the OSIL Technology.

Requirement, Generation and Transfer of Heat in a Sponge Iron Rotary Kiln

7

CHAPTER

7.1 MATERIAL AND ENERGY BALANCE IN A CONVENTIONAL ROTARY KILN SPONGE IRON MAKING PROCESS

We know that whatever material or energy we put into a continuously operating reactor must come out in some or the other form. In a continuous and steady process, there cannot be any accumulation in the process. When one attempts to list the inputs and outputs of both material and energy, a greater understanding of the process is achieved. The same is true of the rotary kiln process of sponge iron making.

A typical material balance of a commercial rotary kiln sponge iron plant is presented in Fig. 2.5. It can be seen that while about 2.5 tonnes of solids are charged, only about 1.4 tonnes of solids are discharged. About 1 tonne of solids are gasified in the process and the balance is carried away as dust in the waste gas stream.

A typical energy balance of a commercial rotary kiln sponge iron plant is presented as a Sankey Diagram in Fig. 2.6. It is disturbing to see that out of 6 GCal energy pumped into the process (entirely as chemical energy in coal), more than 2 GCal or over 37% is lost in waste gases. Many units having multiple kilns use the gases to generate electricity and sell to local grids. While that is a very welcome by-product, the new entrepreneurs who want to make a modest start with, say, a 25 or 50 tpd single module plant is placed at a great disadvantage. The situation calls for evolving means to plough back the waste gas energy into the sponge iron making process itself.

7.2 GENERATION AND TRANSFER OF HEAT

In a sponge iron rotary kiln, the overall reduction process is endothermic and tends to lower the bed temperature.

$$Fe_2O_3 + 3/2\ C = 2Fe + 3/2\ CO_2$$
$$\Delta H^{o}_{298} = +55{,}225\ \text{kCal/kg-mole}$$

(Reaction no. 6 in Table 3.1— '+' sign indicates that heat is absorbed)

To make up for the lost heat the free-board gases are maintained at a substantially higher temperature; and the heat transferred out of the gases to the bed and refractory is made up through controlled combustion of CO in free board gas to CO_2.

$$2CO + O_2 = 2CO_2$$

$$\Delta H^o_{298} = -135{,}260 \text{ kCal/kg-mole}$$

(Difference of reactions numbered 3 & 4 in Table 3.1; '–' sign indicates that heat is evolved)

This reaction is also written as

$$CO + ½ O_2 = CO_2$$

$$\Delta H^o_{298} = -67{,}630 \text{ kCal/kg-mole}$$

Heat from free board gas reaches the granular solid bed in both direct and indirect manner. Since the gases "see" the charge bed only about 25% of the time, the bulk of heat transfer takes place through a via-medium i.e., the refractory surface. The heated refractory surface, in turn, transfers the heat to the solid charge while it goes under the solid bed and they come in intimate contact during rotation.

There is also transfer of heat through refractory on to the kiln surface and to the outside atmosphere. This heat is a loss to the process and should be kept to a minimum with suitable insulating layers in the refractory.

7.3 MATHEMATICAL TREATMENT OF HEAT TRANSFER

Heat transfer in rotary kiln takes place through all possible modes viz., conduction, convection and radiation. While both conductive and convective heat transfers are linearly dependent on the temperature difference, the radiative transfer is dependent on the fourth power of the absolute temperature. For mathematical treatment, nine heat transfer terms can be defined which are pictorially represented in Fig. 7.1. These terms are:

H_1 = Conductive and convective heat transfer from gas to charge bed.

H_2 = Radiative heat transfer from gas to charge bed.

H_3 = Conductive and convective heat transfer from gas to refractory.

H_4 = Radiative heat transfer from gas to refractory.

H_5 = Conductive and convective heat transfer from refractory to charge bed.

H_6 = Radiative heat transfer from refractory to charge bed.

H_7 = Conductive heat transfer through refractory to kiln surface.

H_8 = Convective heat transfer from kiln surface to atmosphere.

H_9 = Radiative heat transfer from kiln surface to surroundings.

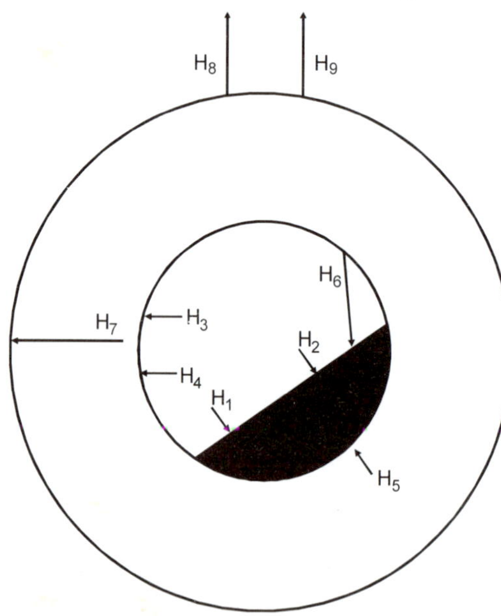

Fig. 7.1 Modes of heat transfer in a rotary kiln

These nine heat flow terms can be expressed in terms of heat transfer co-efficients as follows:

$$H_1 = A_1 (T_g - T_b),$$

$$H_2 = A_2 (T_g^4 - T_b^4)$$

$$H_3 = A_3 (T_g - T_r)$$

$$H_4 = A_4 (T_g^4 - T_r^4)$$

$$H_5 = A_5 (T_r - T_b)$$

$$H_6 = A_6 (T_r^4 - T_b^4)$$

$$H_7 = A_7 (T_r - T_s)$$

$$H_8 = A_8 (T_s - T_a)$$

$$H_9 = A_9 (T_s^4 - T_a^4)$$

Here T refers to absolute temperature and the subscripts stand for the following:

g = gas

b = charge bed

r = refractory surface

s = kiln surface

a = atmosphere/surroundings

Coefficients A_1 to A_9 are the relevant heat transfer coefficients. Temperatures especially in the reduction zone are in such a range that the radiative heat transfer coefficients command a higher weightage than heat transfer by other modes except for the A_5 term, which is the conductive heat transfer term for the heat transferred from refractory to the granular solid bed. Schnabel (1977), based on his own heat transfer model, predicted that during the 108° arc of contact between refractory and the solid charge, refractory loses about 80°C temperature, which it regains during the next 252° exposure to radiative and convective heat transfer from the freeboard gas to the refractory. This represents almost two and a half times higher rate of heat flux (rate of flow of heat per unit area) from refractory surface to solid bed as compared to the heat flux from the freeboard gas to the refractory surface. The higher heat flux is made possible due to intimate interaction and close contact between refractory surface and the solid charge bed as it goes under the latter during kiln rotation.

Based on empirical determination of coefficients by earlier investigators and the knowledge of refractory properties, the heat transfer coefficients A_1 to A_9 can be evaluated and the heat flux can be estimated. Such an estimate would be reliable provided we have reliable estimates of T terms. Measurement of T_g is often tricky as radiation from sensor tip to the refractory surfaces in the range of vision (always cooler than gas), tends to record a lower than actual value. Only a suction pyrometer is likely to record an accurate gas temperature.

Kiln operators, in fact, hardly ever need to know the actual gas temperature. They adjust the various air valves in central burner, shell mounted air fans etc. to maintain a desired level of solid bed temperature. But, as and when accretion formation starts, operators do want to control the gas temperature at the location of accretion to prevent rapid growth. It is a good idea to be guided by the temperature measured by Quick Response Thermocouple (QRT), not because QRT records gas temperatures accurately, but its indication is closer to refractory surface temperature, and it is the refractory surface temperature which is more important in controlling the growth of accretion.

7.4 SEGMENT-WISE AIR REQUIREMENT

It is often a matter of great debate as to what is the standard amount of air required to be put in through the central burner and the various shell air fans. Operators keep on adjusting these flows in order to get a desired temperature profile in the kiln. But as and when the process goes out of control, the blame very often goes towards not maintaining a standardised air flow pattern.

Figure 7.2, which is nothing but a spreadsheet (Excel) chart, represents a simplified method of theoretically estimating

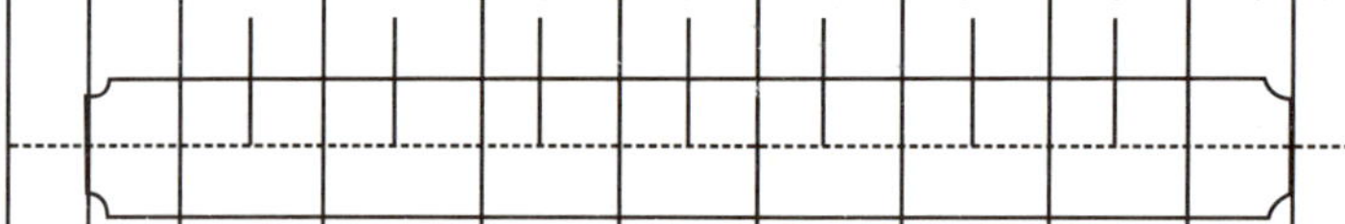

Segment Number	1	2	3	4	5	6	7	8	9	10	11
Solid Bed Temperature [Measured or estimated). °C	900	1000	1000	1000	1000	1000	950	850	700	400	0
Degree of reduction (Interpolated), %	95	95	90	85	80	70	50	30	10	5	0
Increase in Degree of reduction over previous segment, %	0	5	5	5	10	20	20	20	5	5	0
Heat required for reduction of iron ore. GCal/hr	0.0000	0.1294	0.1294	0.1294	0.2589	0.5177	0.5177	0.5177	0.1294	0.1294	0.0000
Heat required for heating iron ore and flux, GCal/hr	0	0	0	0	0	0.0000	0.1766	0.2649	0.5249	0.7065	0
Heat required for heating feed coal, GCal/hr	0	0.0000	0.0000	0.0000	0.0000	0.0367	0.0734	0.1101	0.2202	0.4256	0
Heat required for heating thrown coal. GCal/hr	0	0	0.07448	0.2341	0.3618	0.2873	0.1011	0	0	0	0
Allocation of radiation and unaccounted losses. GCal/hr	0.2659	0.2659	0.2659	0.2659	0.2659	0.2659	0.2659	0.2659	0.2659	0.2659	0.2659
	0.2659	0.3953	0.4698	0.6294	0.8865	0.1959	0.1348	0.1587	0.1455	0.5274	0.2659
Air required in the segment, Nm^3/hr	568.1	844.6	1003.7	1344.7	1893.9	2554.9	2424.2	2475.4	2447.1	3263.1	568.1

Assumptions:
Iron ore rate = 7.5 tph;
Fe = 64%;
Coal rate = 5.6 tph;
Feed coal = 5.6 tph;
Thrown coal = 2.8 tph,
Fixed carbon in coal = 2.8 tph,
Thrown coal = 2.8 tph,
Fixed carbon in coal = 39% (Wet basis);
VM = 26%; Ash = 30%;
Fixed carbon/Total Iron = 0.455;
of the 2.8 tph thrown coal, the following distribution is assumed:

Segment 3 - 7%
Segment 4 - 22%
Segment 5 - 34%
Segment 6 - 27%
Segment 7 - 10%
Flux rate 350 kg/hr; Heat required for full Reduction (100%):
$Fe_2O_3 + 3/2C = 2Fe + \frac{3}{2} CO_2$;
ΔH = 55, 225 kCal/kg-mole

Fig. 7.2 A simplified method of calculating requirement of air in different kiln segments

the air requirements in the various segments of the kiln. Portions before and after the kiln has also been considered as segments, as some portion of the process continues there. In this example, the process has been divided into eleven segments. Nine of these are the sections of the kiln itself, based mostly on the location of the air tubes. Segment number 1 comprises the product discharge transfer chute and the cooler, while segment 11 comprises the feed pipe, the inlet hood and the ABC chamber.

In chapter 5 we have seen a method of determining the total air requirement of the process. Figure 7.2 demonstrates how we can allocate the requirement to the various segments. In the current example, air requirements of segments 1, 2 and 3 have to be met by the air put in through the central burner and the coal thrower. In practice it has been observed that all the air put in by the central burner and the coal thrower are not utilised in these segments. As a rough estimate we can say that only 70% of air is utilised and the remaining amount is utilised in the subsequent segments. The same may be happening in the subsequent segments as well in the reduction zone. In the preheating zone, the air requirement increases rapidly, the gas volume increases and the flow rate becomes highly turbulent. The utilisation factor should then be higher although an assessment has not been made so far.

There is possibility of extending this type of analysis for theoretically estimating the average level of gas and refractory surface temperatures in the different segments. Refractory temperature in particular is an indicative measure of proneness to accretion formation and growth.

8

CHAPTER

Reaction Kinetics

The thermodynamic properties, especially the free energy change of a reaction, establishes whether a particular reaction is feasible or not, but it does not indicate at what rate the reaction would take place. To enhance the rates of chemical reaction, increasingly higher temperatures have been employed over the years, as in the case of extraction of iron, and it has resulted in greater success of the indirect processes over the more direct sponge iron making processes. But persistent attempt at enhancing the rate of solid state reduction process has led to re-emergence and commercial success of the lower temperature and solid state reduction processes, such as the sponge iron making process. Although the attempts at enhancing the rate of solid state reduction has largely been by trial and error, but the science of the study of Chemical Reaction Kinetics (or simply Reaction Kinetics) has been a significant contributor. For further enhancement of rates, Reaction Kinetics is expected to play a key role in future.

8.1 FACTORS AFFECTING REACTION RATES

As implied earlier, the science of Reaction Kinetics studies rates of chemical reaction. Chemical reaction rates (meaning mass of reactants transformed per unit time) generally depend on the following:

- Concentrations (or more accurately the thermodynamic activities) of reactants
- Temperature of reaction – Increase in temperature increases reaction rate
- Physical state of reactants – whether in a gaseous phase or in a solution (homogeneous reactions) or in separate phases (heterogeneous reaction as in a sponge iron rotary kiln) – whether in solid blocks (low surface area) or in powder, colloidal or emulsion form (high surface area). Higher the surface area, higher are the reaction sites and therefore higher are the reaction rates.
- Presence of Catalyst or Inhibitor
- Means of excitation – Light, Ultraviolet or Radioactive Radiation, etc.

8.2 RATE LAW AND ORDER OF REACTION

The most simple reaction is where one constituent breaks up into a number of product constituents; e.g., A = B + C + D +....

If there is no appreciable reverse reaction then rate of reaction is represented by –d[A]/dt, where parentheses '[]' represent concentration of the constituent A. The sign is negative because the concentration of A is decreasing.

'Rate Law' states that the rate of reaction is proportional to the concentration of the reactant(s). Hence,

$$-\ d[A]/dt = k[A]$$

k is referred to as the Velocity Constant of Reaction

or, $$d[A]/[A] = -\ kdt$$

Integrating, we get

$$\ln [A] = -\ kt + C$$

If, in the beginning (i.e., when t=0), [A] $= [A]_0$

then, $$C = \ln [A]_0$$

and $$\ln([A]/[A]_0) = -kt$$

or $$[A]/[A]_0 = e^{-kt}$$

If, in time t, x amount of A has reacted, then,

$$[A] = [A]_0 - x$$

and $$[A]_0 - x = [A]_0\ e^{-kt}$$

or $$x = [A]_0\ (1 - e^{-kt})$$

On the other hand,

$$k = (1/t) \ln ([A]_0/[A])$$

$$= -\ (2.303/t) \log \{([A]_0 - x)[A]_0\}$$

$$= -\ (2.303/t) \log\ (1 - x / [A]_0)$$

A plot of $-\log (1 - x / [A]_0)$ against time t would be a straight line with a slope of k/2.303 from which value of k can be calculated. This particular relationship is said to hold for what is called a "First Order Reaction."

8.3 BIRTH OF THE RATE LAW OR LAW OF MASS ACTION

In 1799 Proust propounded the Law of Constant Proportions which stated that 'elements combine in definite proportions by weight and the composition of a pure chemical compound thus formed is independent of the method by which it is prepared.' Bertholet opposed this law bitterly and spent next nearly sixty year trying to bring evidences against it only to be rebutted one by one. In his quest for gathering such evidences it finally dawned on him and his co-workers that he had been bringing up evidences of partial combination and the extent of combination, in some way, depended on the concentrations of the reactants used. From these results, Guldberg and Waage in 1864 deduced the Law of Mass Action, also known as the Rate Law. This law stated that "the rate at which a substance reacts is proportional to its active mass," and "the velocity of a chemical reaction is proportional to the product of the 'active masses' of the reactants." The active mass was assumed to be proportional to the molar concentration, as has been used in derivation in the last section.

8.4 UNIMOLECULAR REACTION AND FIRST ORDER REACTION

In the deviation of section 8.2 we have taken the example of a unimolecular reaction, i.e., a single molecule of A breaks into a number of molecules. We have also said that the final derivation

$$-\log(1 - x/[A]_0) = (k/2.303)t$$

is applicable to a first order reaction.

By a similar logic, for a bimolecular reaction of the type A + B = C + D + ...

and starting with equal concentrations of A and B (i.e., $[A]_0 = [B]_0$) we get the relationship

$$1/(1 - x/[A]_0) = 1 + akt$$

and this applies to a second order reaction.

Although the concept of 'Order of Reaction' has evolved from the molecularity of reaction, but they need not be the same. While molecularity refers to the number of molecules involved in a chemical reaction, the order of reaction must be derived from the study of the variation of reaction rates with time, concentration and (as we shall see later) temperature.

8.5 DECOMPOSITION OF AN OXIDE

Let us consider a chemical reaction where an oxide breaks into its elements.

$$MO = M + ½O_2$$

Degree of reduction, in such a case, is defined by the ratio of "oxygen removed from the oxide" to the "total removable oxygen." and is represented by the symbol α. Thus

$$\alpha = 1 - [MO]/[MO]_0$$

and a plot of $-\log(1 - \alpha)$ with t at any given temperatures should give a straight line with a slope of k/2.303 from which the value of velocity constant of reaction at that temperature can be calculated. This way the reaction is found to be of the first order. But if the plot is not a straight line, the reaction is not of first order, even though it remains a

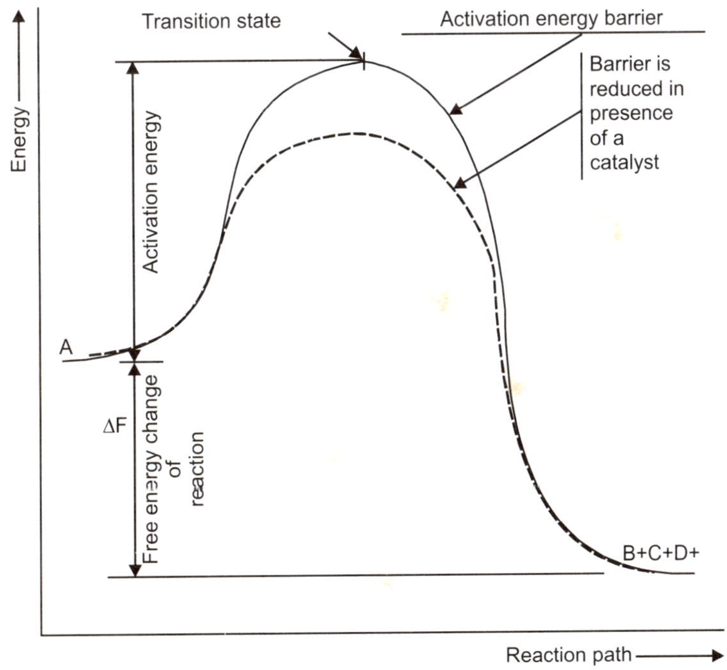

Fig. 8.1 Schematic diagram representing activation energy barrier and the effect of a catalyst

unimolecular reaction. Very often the reactions take place in a number of steps and are much more complex than what they appear at first glance. We shall see later that the reduction of iron oxide to metallic iron in a rotary kiln is also highly complex and the actual mechanism of reaction can only be conjectured at the moment.

8.5.1 Degree of Reduction

We have defined degree of reduction in the last section. If we take that definition literally for different types of iron oxides (such as hematite and magnetite) it would become highly difficult to compare results from different sets of experiments.

For circumventing this problem, we define $\alpha = 0$ for that iron oxide where all iron atoms are present as Fe_2O_3 molecules. If we have to start a study with a sample of say magnetite, then we say that the sample, to start with, is already 11.11% reduced and $\alpha = 0.1111$. It then becomes easier to compare results of different studies while using such a basis for defining degree of reduction.

8.6 TEMPERATURE DEPENDENCE OF VELOCITY CONSTANT: THE ARRHENIUS EQUATION

We have so far dealt with velocity constants of reactions at a particular temperature. In 1887 Arrhenius postulated that the velocity constant k of chemical reactions change with temperature as per the relationship

$$k = Ce^{-E/RT}$$

E is referred as the activation energy and it can be visualised as a barrier (an energy barrier) to the chemical reaction (Fig.8.1). Reacting molecules must acquire, during random collisions, interactions or excitation, enough energy so that it is able to cross the dam-like barrier and form the product. An increase in temperature would mean that molecules in general have acquired higher energy and there is an increased probability of greater number of molecules crossing that barrier in a given time.

Thus

$$\ln k = C' - E/RT$$

or, $$\log k = C'' - (E/4.576)(1/T)$$

where R and E are expressed in calories. Thus, for evaluating E, we need to plot log k against 1/T and measure the slope of the resulting straight line.

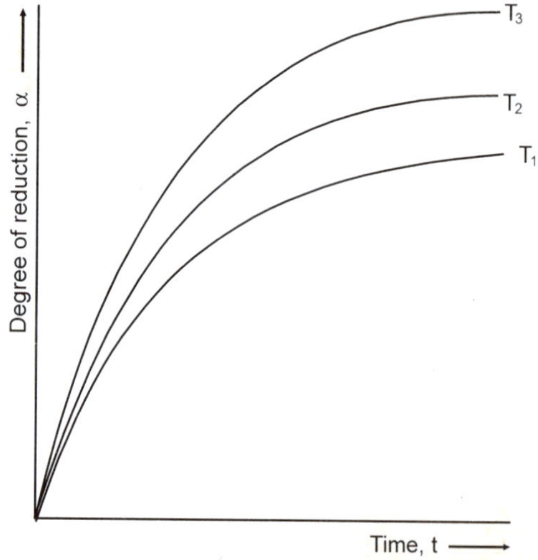

Fig. 8.2 Typical plots of degree of reduction with time at different temperatures

8.7 EXPERIMENTAL DETERMINATION OF ACTIVATION ENERGY

In case of decomposition of an oxide as described in section 8.5, plots of degree of reduction against time would be of the type given in Fig. 8.2. These plots for different temperatures are experimentally determined. Various experimental techniques are followed to obtain the desired results, but Thermogravimetry is a quick and convenient method if adequate precautions are taken.

Once these results are obtained then $-\log(1-\alpha)$ values are plotted against time t, and they should be straight lines if the reaction is of first order. If reasonable linearity is observed then their slopes are measured (Fig. 8.3) to obtain k or velocity constant values. Thereafter, log k values are plotted against 1/T (reciprocal of absolute temperatures), and slope of this plot is a measure of activation energy. In fact, in caloric terms, the slope is equal to $-E/4.576$, where E is the activation energy (Fig. 8.4).

If plots in Fig. 8.3 are not straight lines the reaction is not first order. For second order reaction, instead of $-\log(1-\alpha)$, a different function of α needs to be plotted along the Y-axis. This function is $1/(1-\alpha)$. This time the plots would be straight lines, but not passing through the origin.

Instead of trying to find out the order of reaction by trial and error, a short cut is normally applied by using what is called a reduced time plot. In this plot α is plotted against a dimensionless time, which is obtained by taking ratio of the actual time with the time required to attain $\alpha = 0.5$. On the plot obtained experimentally, theoretical plots of reaction by different mechanisms are super-imposed and then the matching with a particular mechanism can be quickly observed. Many times in a particular reduction region, one mechanism may be followed, which would be giving way to another mechanism in another region, and this also can be deduced by comparing with standard plots (Fig. 8.5).

8.8 VARIATION OF REACTION RATE WITH TEMPERATURE

The treatment presented in the proceeding sections is applicable to the rates of reactions. First order and second order reactions have been discussed. Reaction can be of higher order (third order and above) as also of fractional order. Reactions can also be of zero order where there is no effect of concentration on reaction rate. But we hardly ever come across such a reaction.

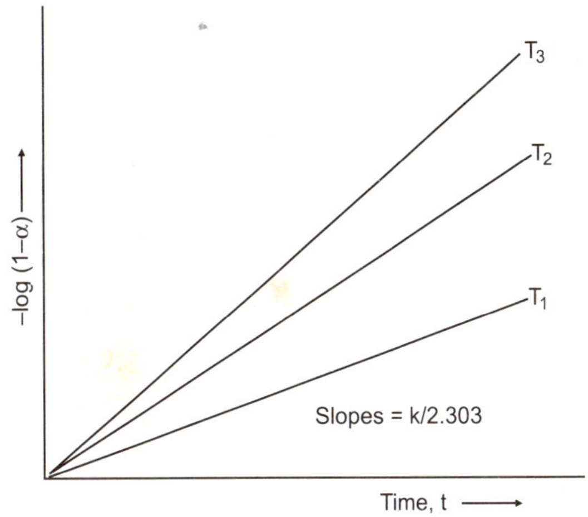

Fig. 8.3 Plot of $-\log(1-\alpha)$ with time (Straight Lines denote First Order Reaction)

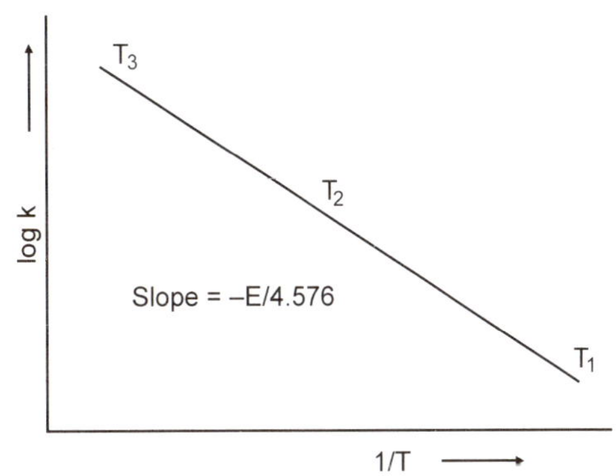

Fig. 8.4 Plot of log of velocity constant against reciprocal of absolute temperature (Arrhenius plot)

There is a thumb-rule often quoted for Metallurgical Reactions, according to which if we increase temperature by 10°C, chemical reaction rates get doubled. It is very unclear as to how this thumb-rule came into prominence. On analysing the variation in velocity constant at about 1000°C we find that the above reasoning holds only if the activation energy is of the order of 220 kCal/mole.

Let us take the example of iron oxide reduction in solid state by carbon monoxide, as it happens in the sponge iron making process.

$$FeO + CO = Fe + CO_2$$

Even though it is a bimolecular reaction, a large majority of researchers who investigated it experimentally, have reported it to be of first order. The activation energy values though, vary very widely, and are much lower than that expected from other similar reactions. It is reasoned that the metallic iron which initially forms, catalyses and therefore enhances further reaction. If the activation energy value is in the range of about 55 to 60 kCal/mole, then the reduction rate would be enhanced by only 20% when temperature is increased by 10°C.

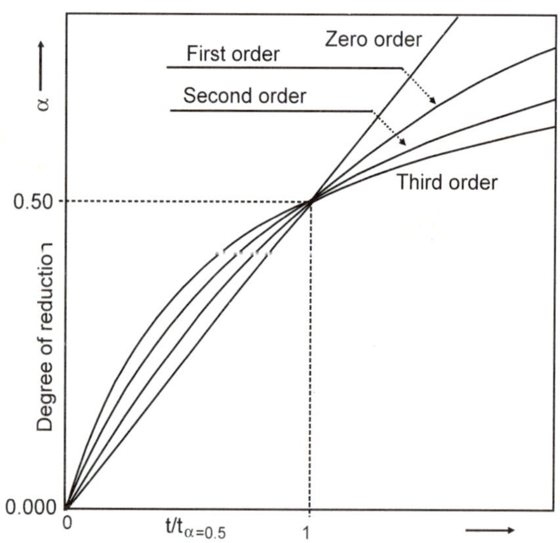

Fig. 8.5 Reduced time plots for different orders of reaction

8.9 ROLE OF DIFFUSION

In case of a gas-solid reaction, as in the example cited in the last section on sponge iron making, the reaction product, if solid, tends to form a barrier in the interaction process. The gaseous species has to diffuse through the product layer in order that further reaction is continued. Alternatively, the solid reactant may dissolve in the product layer, or in some other way diffuse through the product layer to interact with the gas phase. If these diffusion rates are sufficiently fast, the reactions may still proceed, with the chemical reaction rate being the controlling step, and activation energy of the chemical reaction would determine the variation of reaction rate with temperature. This probably happens in the sponge iron making process as a majority of researchers have reported this process of sponge iron making to be "Chemical Reaction Rate Controlled."

But if diffusion rates are too slow the activation energy of chemical reaction would not have any role in the progress of reaction and we would have a situation where the progress of reaction is 'Diffusion Controlled.' In intervening regions where both chemical reaction rate and rates of diffusion have influence on overall progress of reaction we say that the reaction progress is in the 'Mixed Control' domain.

We would not discuss here the diffusion mechanisms and diffusion controlled processes. The treatment is complex and other comprehensive documents may be referred for this purpose. We would only emphasise the need to find means to circumvent diffusion steps in reactions involving condensed phases (either solids or liquids) as diffusion barriers often make industrial exploitation of process unviable.

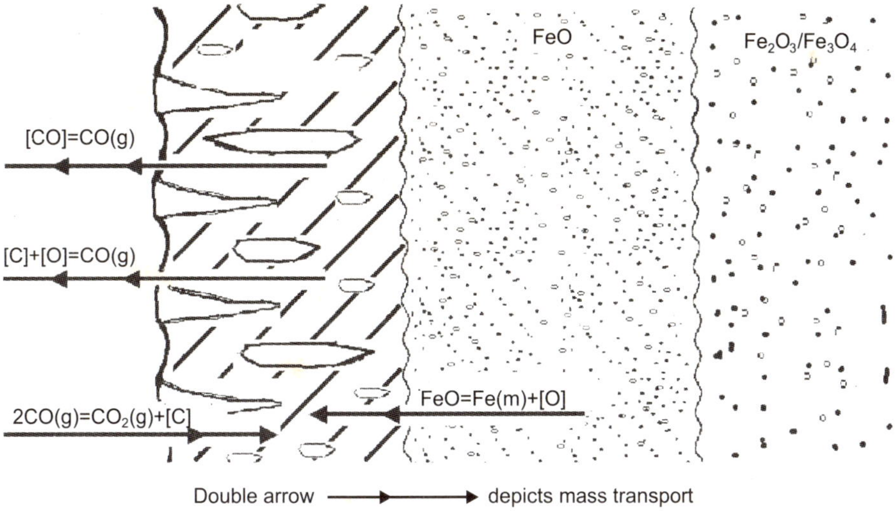

Fig. 8.6 A possible mechanism of pore formation during reduction of iron oxide

8.10 HOW SPONGE IRON MAKING IS A FIRST ORDER REACTION?

As mentioned earlier that although the following reaction

$$FeO + CO = Fe + CO_2$$

is bimolecular, a very large number of researchers have found it to be of first order. A few reasons may be responsible for this observation.

- FeO being in condensed phase, its activity (representing concentration) does not change with time and therefore only the concentration of CO is affecting the rate of reduction.
- The reaction is, in some way, being catalysed. Many say that the catalysis is by the iron metal, which is produced in the reaction. During catalysis, an intermediate single molecule is formed, which then breaks into the products. Therefore, the reaction is of the first order.

Largely following the second reasoning, a hypothesis on the mechanism of reduction has been presented in Fig. 8.6. Since there is large volume reduction (over 50%) during sponge iron making, a lot of pores form on the product layer, which is formed by the reduction on surface of iron ore lump or pellet. While these pores facilitate easier access to reducing gases, it is not necessary for CO molecules to actually meet FeO molecules to effect further reduction. Carbon and oxygen have significant solubilities in metallic iron. The dissolved carbon and oxygen atoms within the metal layer probably form an associated molecule (CO dissolved in metallic iron), which get liberated on the surface to give CO gas and thereby effecting the reduction of FeO.

$$2CO(g) = CO_2(g) + [C]$$

Here, parentheses '[]' mean that the constituent is dissolved in Fe_m layer.

$$FeO(s) = [Fe] + [O]$$

$$x[C] + x[O] = [CO]_x$$

$$[CO]_x = x\,CO(g)$$

It is open to future researchers to investigate and prove or disprove this hypothesis. At the time of writing it appears to be a plausible hypothesis.

Raw Materials for Sponge Iron Making

CHAPTER 9

Iron Ore, or more strictly the iron oxide feed, is the source of iron in sponge iron making. Coal is the other major raw material which is used both as a fuel and the reductant in this process. The third raw material is the flux, either limestone or dolomite, which is used to desulphurise iron ore and to prevent sulphur in coal from entering in sponge iron. Since flux is used in only minor quantities, and its quality has negligible influence on the process, it will not be discussed here.

9.1 IRON OXIDE FEED

India is fortunate enough to be possessing large quantities of high grade iron ore deposits. These are mined, sized and used directly in sponge iron kilns as well as in other metallurgical reactors like gas based sponge iron shaft furnaces and iron making blast furnaces. The other iron oxide feed which is used in rotary kiln is the iron ore pellets. Pellets are formed by what is called the "balling" when finely ground iron ore mixed with small quantities of moisture and suitable binders is rotated or rolled over the surface of a rotating disc or drum generally along with slight sprinkling of water. These pellets are "green" and weak and are fixed in furnace in excess of 1200°C for it to harden. Till recently the higher cost of pellets prevented its economic use in rotary kiln, but with reduced availability and higher cost of quality lump ore, coupled with the understanding that pellets improve kiln productivity, pellets are now becoming a viable alternative.

Unlike pellets, sinter cannot be used in sponge iron making. Sinter is a very suitable feed for blast furnace iron making as the emphasis there is on melting rather than solid state reduction. Since sinter is formed by surface fusion, the iron oxide structure and properties have changed and it is not amenable to solid state reduction. Iron ore briquettes, formed by cold compaction of iron ore fines along with a binder, have sometimes been tried, but without much success. During reduction, briquettes tend to decrepitate or break up resulting in production of large quantities of fine sponge iron of reduced value. Also when the kiln carries a large load of fines, accretion or ring or deposit formation is accelerated.

9.2 IRON ORE LUMPS

Almost all high grade iron ore deposits of India contain Hematite (Haematite Fe_2O_3)

mineral. These are associated with undesirable (gangue) minerals containing Silica SiO_2, Alumina Al_2O_3, etc. Pure hematite has the chemical formula Fe_2O_3 and therefore contains 69.94% iron, balance 30.06% being oxygen. Presently, sponge iron plants are able to get iron ore lumps containing about 64% iron.

From the ore deposits, ore is taken out by removing the overburden, blasting the iron ore seams, excavating and crushing and sizing. While the sized ore is consumed by iron making units, ore fines left at mine site are not usable directly and needs to be beneficiated and/or agglomerated. Further, during mining, seams of lower grade material are also excavated, which do not find ready consumers and are often kept aside for a possible future treatment by beneficiation and agglomeration (mostly pelletisation).

9.3 MAGNETIC AND HEMATITIC ORES

We have so far talked mainly about hematite (also spelled haematite) ore deposits. But in nature magnetite or Fe_3O_4 formed first. Hot liquid volcanic eruptions brought out liquid iron silicate at some locations. Silica being higher melting, solidified into a silica rich quartz layer. Solidification of silica resulted in enrichment of adjoining liquid layer in iron oxide which latter solidified into magnetite, being the stable compound at that temperature, i.e., near 1500° C. The process resulted in what is called Banded Iron Formation (or BIF), having layers of quartz and magnetite of varying thickness. When the layer thicknesses were small, Banded Magnetite Quartzite (BMQ) was formed. Thus, most of the magnetite deposits and BMQ's are of igneous nature (i.e., formed directly from volcanic eruptions) and have relatively dense structure. Natural action of heat, air and water has caused oxidation and re-crystallisation of some magnetite deposits and also the magnetite in BMQ into hematite. In this process we get what is called metamorphosed hematite and in the latter case, Banded Hematite Quartzite (BHQ). Such hematite deposits have relatively open structure and are more easily reducible than magnetite. However, BHQ remains heavily contaminated with silica. But freaks of nature has resulted in natural beneficiation of such deposits. Action of heat, acidic water, air and sun has resulted in slow dissolution of this iron oxide, its collections in ponds and lakes formed in volcanic craters nearby and slow but massive precipitation through formation of sediments. These deposits are called sedimentary type and most of the high grade hematite deposits are of this type. Since the deposit has formed by low temperature precipitation, it has an open and porous structure and is therefore highly reducible. At some places, these deposits have been enriched further by the action of basic water which leaches away silica. But removal of silica weakens the bonding of deposit and we get a friable deposit, sometimes even blue dust.

The Gorumahisani Iron Ore Mines is an example where both igneous and sedimentary iron ore occur in close proximity. Sedimentary lakes had formed within the crater of the Gorumahisani Hills.

9.4 MAGNETIC ORE MINERALS

Magnetite or Fe_3O_4, derives its name because of its magnetic property. Its magnetic intensity is about 40% that of pure iron. Since most magnetite deposits are igneous and difficult to reduce, sponge iron producers avoid using it in rotary kiln. But being magnetic does not always mean the ore is igneous type. Sometimes, hematite marginally transforms into its magnetic forms, martite and maghemite. Since these are formed in sedimentary deposits, they retain the open

Fig. 9.1 A view of the Gorumahisani iron ore mines (Shows high grade rocks associated with lower grade and clayey overburden)
Courtesy: M/s Ghanshyam Mishra & Sons Pvt. Ltd.

structure and can be as reducible as hematite in rotary kiln.

9.5 LABORATORY TEST ON (METHOD OF DETERMINING) IRON ORE REDUCIBILITY

We have talked of more and less reducible iron ore in general terms. But for the purpose of quantitative assessment, whether a particular source of iron oxide is more or less reducible than the other, a more precise definition is required.

Reducibility refers to the ease of removal of oxygen, which is combined with iron in the iron oxide feed.

As the reaction process is gas solid interaction process, reduction in the interior of lumps and pellets should have been very difficult, had it not been accompanied by a large volume reduction (52%). The resulting pores which form on the surface allow gases to access the unreduced regions of the lump or pellet and thereby makes the reduction commercially viable. As reduction is effected mostly by the regenerated carbon monoxide in the rotary kiln, what is measured in laboratory for assessing its suitability is the reducibility toward carbon monoxide.

As hematite ores can be reduced upto 33% without effecting any metallisation, measuring reduction rates up to 33% reduction region would not serve any purpose as the key step in the process is the reduction of FeO to metallic iron.

Generally, reducibility is measured as the rate of reduction under standardised condition after reducing gases have removed 40% of the oxygen – what would have originally been with iron in Fe_2O_3 form (i.e., at atomic ratio of O/Fe = 0.9). Results are expressed as "percent reduction per minute." In this connection, one may refer to the latest revisions of ISO/TC 102/SC3, ISO/DP 4695, DIN 1770, JIS (M6495) and IS 8167 of BIS standards on Reducibility of iron ore, sinter and pellets.

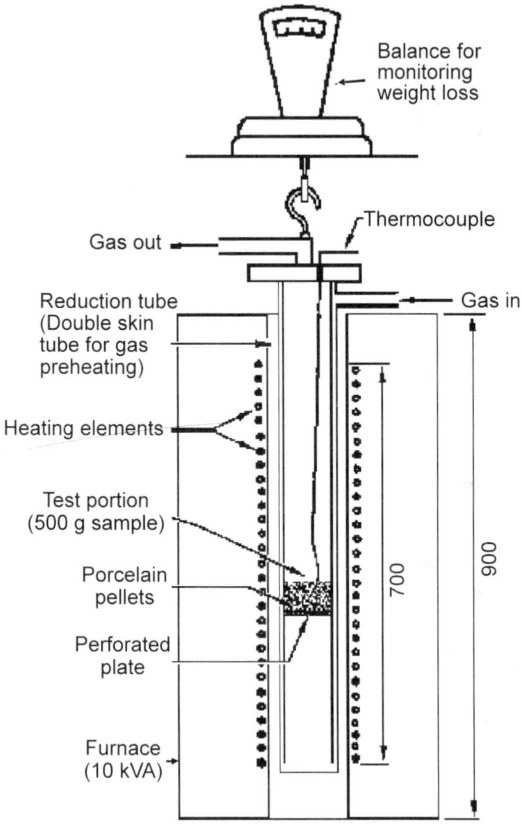

Fig. 9.2 Apparatus for measuring reducibility of iron ore

Almost all high-grade hematite ores have been found to have adequate reducibility for sponge iron making. With higher gangue content, the reducibility generally decreases, probably due to hindrance caused by gangue to the diffusion path of gases. Ores containing large percentage of magnetite are found to have much lower reducibility as compared to hematite ore, possibly due to marginally smaller pore volume formation in the reduced material (50%) and generally higher levels of gangue associated with magnetite, and also, as mentioned earlier, due to relatively denser structure of magnetite as most magnetite deposits are formed from the melts of volcanic eruptions.

9.6 COAL FOR SPONGE IRON MAKING

Coal is a combustible mineral substance (rock) generally black or blackish in colour. Coal is the main source of meeting industrial energy requirement. It is still far ahead of its cleaner alternatives such as petroleum, natural gas, electrical and atomic energy. Coal is called a fossil fuel as it is formed by fossilisation of wood and plant matter over centuries and millennia.

Coal is formed from ancient plant or woody matter initially by biological action (peat) followed by physico-chemical compaction over millions of years. Most of the coal deposits were formed during, what is called, the Carboniferous Period ranging from about 360 to 300 million years ago. This period was highly conducive to growth of plants and trees of gigantic sizes and numbers. But the conditions of rainfall, river movement and shifting of earth's crust resulted in periodic destruction of the rain forests, their submergence or carry over and accumulation of woody matter into large lakes and their subsequent submergence by plate shifting.

9.7 PEAT

By micro-biological degradation the lowest rank of coal i.e., peat, was formed, in which the woody structure is retained. It contains more than 50% free moisture as mined. But on suitable storage and exposure to sun, free moisture can be brought down to below 20 % when it can be used as a slow burning fuel.

9.8 CONVERSION OF PEAT INTO HIGHER RANKS

Whenever, natural forces caused submergence of peat deposits and it was covered with an overburden it was subjected to pressure and temperature (e.g., geothermal heat). It slowly transformed progressively into higher ranks viz., lignite or brown coal, high volatile bituminous, medium volatile bituminous, low volatile bituminous, semi-anthracite and anthracite coals. Progressively the deposits gained in carbon percentage while it lost hydrogen, oxygen, etc. Carbon chains become longer and the coal compounds became like a resin (at medium to low volatile bituminous range) which further broke up when the rank increased.

The above may appear like too much of a detail, but it is at this point (resin like formation) that the coal started becoming unsuitable or less suitable for sponge iron making.

9.9 INDIAN COAL DEPOSITS

Fortunately for sponge iron makers (and less fortunately for blast furnace iron makers), most of the Indian Coal deposits are of lower rank type. On the other hand, rather unfortunately, most of these lower rank deposits have mineral matters well dispersed within their seams. Therefore, they yield large amounts of ash on burning and their value is correspondingly reduced. Mineral matter being well dispersed, the coals are not easily amenable to beneficiation. Mineral matters are liberated only at high levels of grinding. Grinding makes the coal less suitable or even unsuitable for most applications and need to be agglomerated before use. The process of beneficiation of coal thus becomes uneconomical for most applications.

9.10 SIZE OF COAL FOR SPONGE IRON MAKING

For sponge iron making, coal from feed end is required to be roughly of 5 to 20 mm size and the coal from discharge end (thrown coal) can be a judicious mixture of fine coal (0 to 5 mm) and lumpy coal (up to 20 mm). Use of ground coal is presently not considered practical in rotary kiln.

9.11 ROLE OF COAL IN A SPONGE IRON ROTARY KILN

Coal is required to perform two vital functions inside the sponge iron rotary kiln. Firstly, it provides the requisite heat for bringing the reactants to the reduction temperature. Secondly, carbon in the char generated from the de-volatilisation of coal acts as a reductant for converting the iron oxide into metallic iron. Of the iron oxides, FeO or Wüstite is the most difficult to reduce. If coal is able to reduce FeO into metallic iron under rotary kiln conditions, then the coal is suitable for rotary kiln sponge iron making process. We would see later that once the coal is found capable of reducing FeO to metallic iron, it is automatically able to provide the necessary heat for the process.

In sponge iron rotary kiln, FeO is reduced as per the following steps:

$$FeO + CO \leftrightharpoons Fe + CO_2$$

Free energy considerations indicate that for this reaction to proceed in the forward direction under rotary kiln conditions, the concentration (percentage) of carbon monoxide will have to be higher than 2.5 times the concentration (percentage) of carbon dioxide. Thus any carbon dioxide either formed by this reaction, or any other source, has to be removed quickly from the site of reaction. Carbon in char can convert CO_2 to

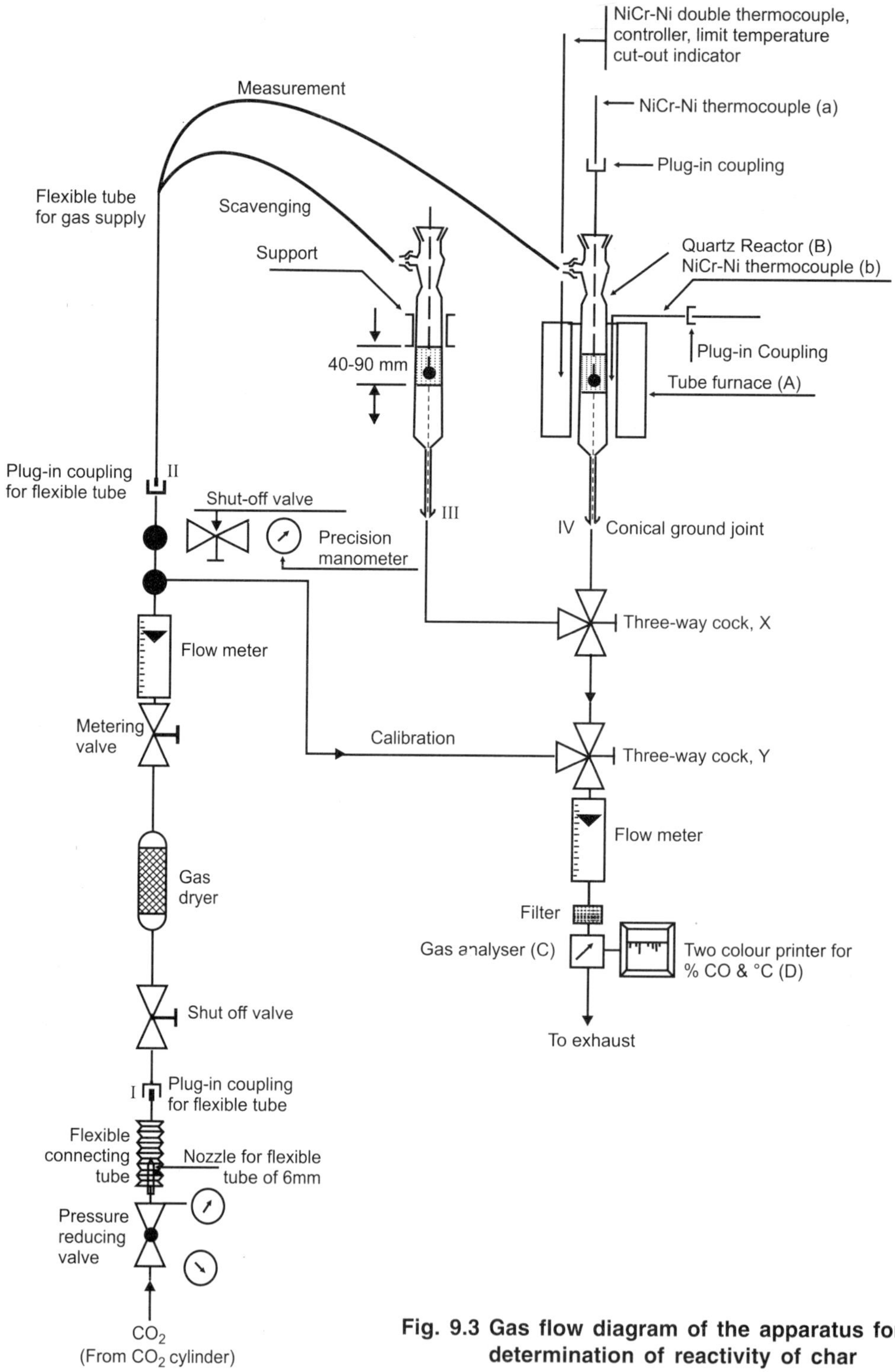

Fig. 9.3 Gas flow diagram of the apparatus for the determination of reactivity of char

CO thus providing additional CO for reduction.

$$CO_2 + C = 2\ CO$$

The more rapidly carbon in char is able to convert CO_2 to CO, the more suitable the coal is for rotary kiln sponge iron making. We say that the char from coal has adequate "reactivity".

We thus define reactivity as the ability of carbon in char to quickly react with CO_2 to form CO. More strictly we are defining "Carboxy-Reactivity" which is relevant for rotary kiln sponge iron making. Other forms of reactivities are "Oxy-Reactivity" – ability of carbon in char to quickly combine with oxygen to form CO – and "Hydroxy-Reactivity" – ability of carbon in char to quickly combine with steam to form CO and H_2.

By combining the reduction and the gasification reactions, we get the overall reduction reaction in rotary kiln sponge iron making.

$$FeO + C = Fe + CO$$

There is thus a surplus of carbon monoxide gas which keeps coming out of the bed in the reduction zone. Partial combustions of this CO supplements at least part of the heat required in the process. Thus we see that once coal is found suitable for the role of reduction, it can automatically fulfil the role of heat provider at least partially.

9.12 MEASUREMENT OF REACTIVITY OF COAL CHAR

For the methods of measurement of the (carboxy-) reactivity of coal char, the latest revision of IS: 12381 may be referred. The apparatus and the procedures mentioned here either measure reactivity in terms of cubic centimetre of CO_2 converted to CO per g. carbon in the bed per second, or measure a relative reactivity in terms of percentage of carbon reacted under specified conditions. Both apparatus and procedures are complex. To further complicate the problems, the results differ with different char preparation methods. The reactivity of char is adversely affected, if temperatures used in its preparation are higher, if it is exposed to maximum temperature for longer period of time, and the results are vitiated if a small amount of VM is retained in the char. The latter often happens if attempt is made to make larger quantity of char in one batch. In future revisions BIS will have to address these aspects in the standard.

9.13 GAS ANALYSIS METHOD OF REACTIVITY MEASUREMENT

We would like to present here briefly the gas analysis method of measuring the reactivity of coal char. IS: 12381 of BIS details two such methods out of which the method involving continuous recording of percent CO in the exit gas using an infra red CO analyser has been selected. Other than the gas flow scheme, the apparatus is quite simple. The gas flow diagram along with the apparatus is represented in Fig. 9.3.

In this method, both instantaneous reactivity and average reactivity over a period can be measured. In all the alternative methods, only average reactivity over a period can be measured. We feel that the results obtained are more reliable as it can easily eliminate the error due to any residual volatile matter in the sample. Other methods can reduce the error on this account by extending measurement period but cannot completely eliminate it.

This method is also free from errors arising out of dropping or blowing off of ash particles.

9.13.1 Principle

For intimate contact carbon dioxide should be passed through a column of char. Since carbon monoxide is much lighter than carbon dioxide, channelling in the column is expected if gases pass up through the column. When gases are passed down the column all the problems mentioned above are overcome, but lower layers of char encounter gas which is getting progressively enriched by carbon monoxide. To account for this phenomenon the concept of Damköhler Number has been incorporated. Damköhler Number represents the ratio of residence time of the gas to the mean life of a reacting species in the gas.

Use of this approach enables us to measure reactivity values instantaneously and the need to average out over a period is obviated.

It is customary to measure reactivity at a point when 0.5g. of carbon has been gasified from a sample of 5g. To determine the instant (i.e., the time) when gasification of 0.5g. of carbon is completed, a graph is drawn at the end of the experiment, between instantaneous gasification rate of carbon (g.C/h) and time. The instantaneous gasification rate of carbon is obtained from the %CO in the exit gas and the gas flow rate. A sample plot is given in Fig. 9.4. The instantaneous gasification rate can be shown to be related to the percent carbon monoxide in the exit gas by the following relationship.

Instantaneous gasification rate

$$= V^*(273/(t_0 + 273))^*(12/22.4)/(200/\text{Vol.\%CO} - 1) \text{ g. carbon/hr.}$$

$$= \text{Vol.\%CO}/(45.27 - 0.228^* \text{Vol.\%CO}) \text{ g.C/hr.}$$

where V = Volume flow rate of CO_2 in litres/hour,

t_0 = room temperature in degrees Celsius

After plotting the graph, the area under the curve is counted to locate the instant (i.e., the time) when the integrated area equals 0.5 g. of carbon removed.

The reactivity may be defined as the instantaneous rate of evolution of carbon monoxide per unit carbon present. The carbon remaining at that instant (when 0.5 g C is removed) is 0.5 g. less than that present in the beginning of the experiment and is therefore equal to (W^*C_{fix} – 0.5) g.

It can be shown that

Reactivity

$$= (V^*(t + 273)/(t_0 + 273)/3.6)^*Da/(W^*C_{fix} - 0.5)$$

Where Da is the Damköhler Number, which is given by

$$Da = 2^*\log_e((200 - \text{Vol.\%CO})/(200 - 2^*\text{Vol.\%CO})) - \text{Vol.\%CO}/(200 - \text{Vol.\%CO})$$

In the present case

$$\text{Reactivity} = (V^*1000/36000)^*(1273/298)^*Da/(5^*C_{fix} - 0.5)$$

$$= 10.68^*Da/(5^*C_{fix} - 0.5)$$

Room temperature has been assumed to be 25°C. For matching results, when conducted at a different room temperature, it is recommended that gas flow may be adjusted using Gas Law to obtain the same expression for Reactivity.

9.13.2 Test Conditions

A sample of 5 g. of char is taken having a size range of 0.5 to 1 mm. The following test conditions are maintained:

a) Temperature : 1000 ± 5°C

b) Test duration : 30 minutes approx.

c) Gas composition : Pure and dry CO_2 gas

d) Gas flow rate : 9 litres/hr.

9.13.3 Test Apparatus

The set up for reactivity determination is represented in Fig. 9.3. This essentially consists of an electrically heated tubular furnace (A), a quartz reactor tube of approximately 15 mm internal diameter (B), an infra red detector for continuous measurement of carbon monoxide in the exit gas (C), and a recorder for recording temperature and %CO in the exit gas (D). The furnace is controlled preferably by a PID controller with an accuracy of ± 5°C. The furnace tube should be 200 mm long and about 25 mm in diameter.

9.13.4 Test Procedure

The reaction tube is constricted in the lower end as shown in Fig. 9.3. A plug of quartz wool is placed above the constriction and is carefully pressed in taking care not to block gas passage.

For measurement of sample temperature, thermocouple tip is placed in the uniform temperature zone of the sample. Thermocouple should be adjusted using a locking ring.

5 g. of char sample having a grain size of 0.5 to 1.0 mm. is filled in the reaction tube and the upper glass tube with the thermocouple placed in position. The tip of the thermocouple should be well within the sample and care should be taken that it is not in contact with the side of the reaction tube.

Gentle tapping should stabilise the filling. The depth of the bed should be 40 to 90 mm. depending on the bulk density of the char. The reaction tube is then mounted on the support outside the furnace and is connected to carbon dioxide cylinder as indicated in Fig. 9.3. Gas is allowed to pass for about 15 minutes (Scavenging) for flushing out all traces of air. Simultaneously, furnace is brought to the set temperature of 1000 ± 5°C. Once scavenging is complete, the reaction tube is removed from the conical ground glass joint III, inserted into the furnace and is connected to a similar ground glass joint IV as indicated in Fig. 9.3. The thermocouple (a) is connected to the printer for continuous sample temperature record.

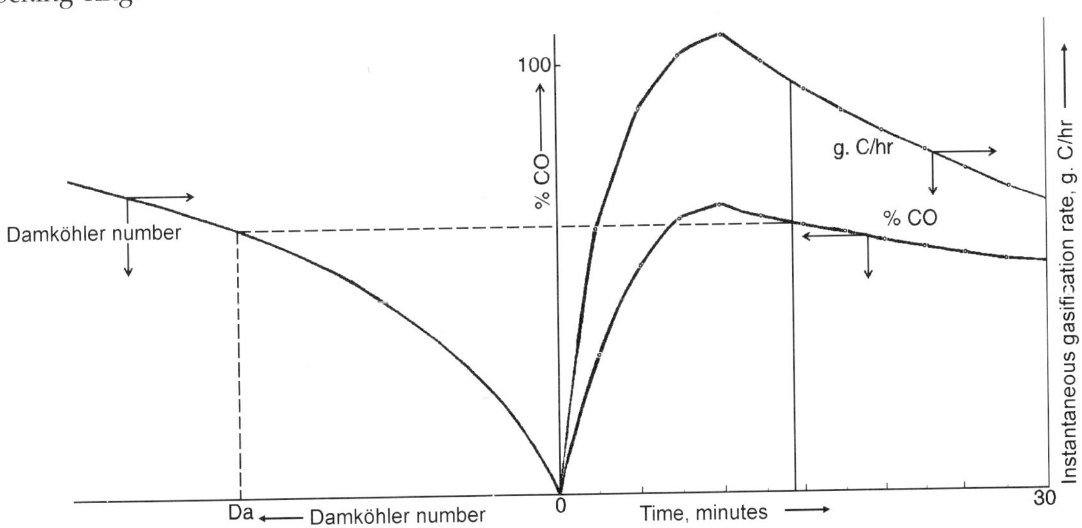

Fig. 9.4 Determination of Damköhler number

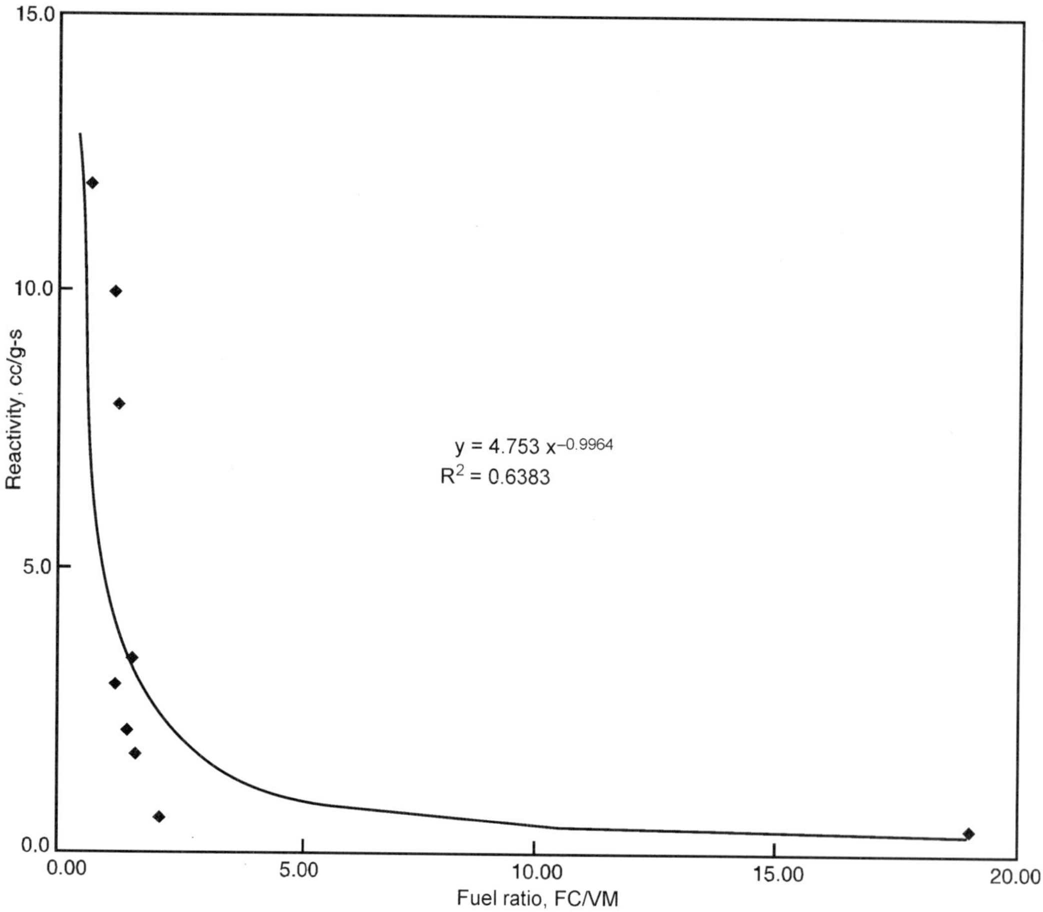

Fig. 9.5 Variation of reactivity with fuel ratio (Data from Reuter & Serbent, 1976)

Printer is started and carbon dioxide gas is passed at the rate of 9 litres per hour (at 25°C) with both three-way cocks 'X' and 'Y' having been set to 'measurement'. After measurements are complete, gas flow is stopped, reaction tube is taken out of the furnace and placed back on the support.

The % CO in the exit gas is plotted against time in minutes (Fig. 9.4). From the values of % CO, initial flow rate of CO_2 (9 litres per minute at 25°C in this case), standard litres of CO evolving per minute can be calculated for each measurement point. This is a measure of grammes of carbon being gasified per hour (g.C/hr), and this is also plotted against time as shown in Fig. 9.4. By measuring area under this plot, that particular time can be located at which point the gasification of 0.5 g. of carbon from the sample is complete. The corresponding per cent CO in the gas is located and this indicates the Damköhler Number.

Reactivity is reported as follows:

Reactivity = $10.68^*Da/(5^*C_{fix} - 0.5)$ c.c./g.-sec.

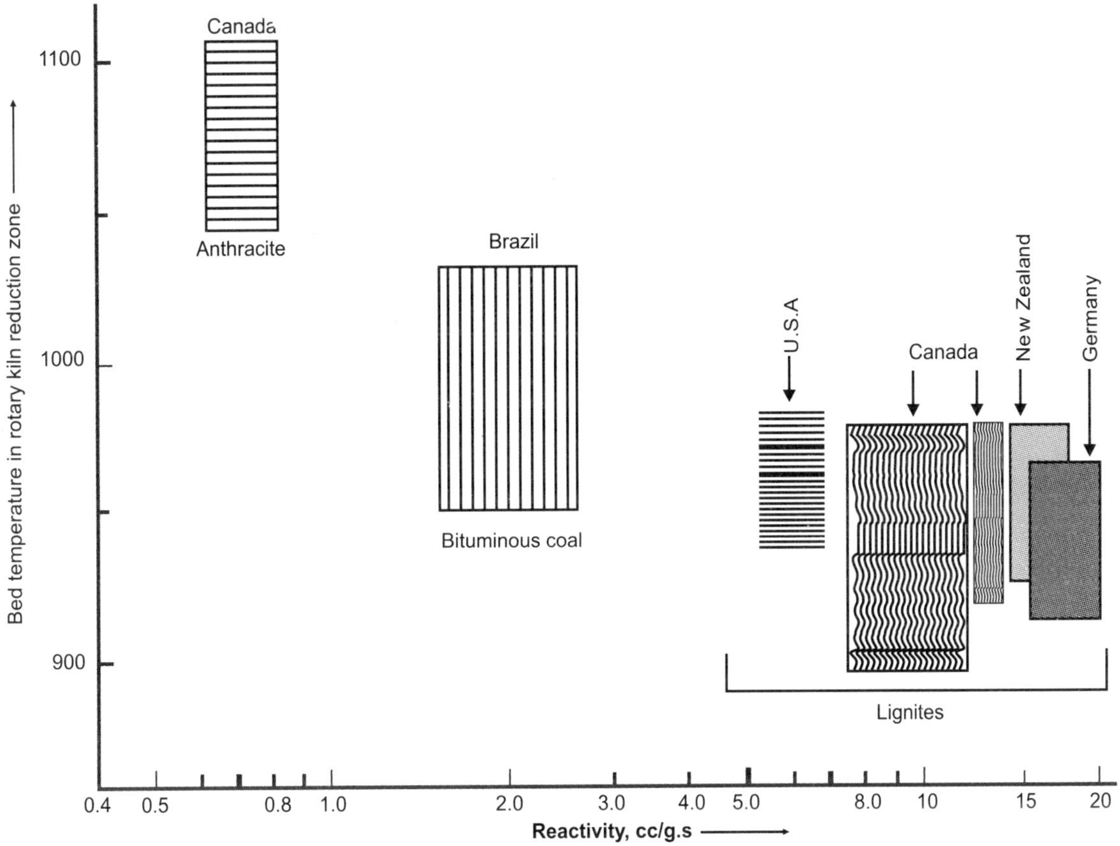

Fig. 9.6 Variation of rotary kiln reduction zone bed temperature with coal char reactivity (Data from Reuter & Serbent, 1976)

9.14 INDICATORS TO REACTIVITY

Since the results of reactivity tests are of doubtful reliability, sponge iron makers rarely go for these tests unless and until they encounter some problems. Although M/s Lurgi, Germany, have made elaborate investigations establishing coal char reactivity to be the most important property for selection of coal for sponge iron making, actual reactivity tests are rarely used in India for this purpose. Some of the important findings of Lurgi are being reproduced here in the form of graphs (Figs. 9.5 and 9.6). Instead operators prefer to use indicators in locating reactive coals keeping the following in mind.

- Lower rank coals have higher reactivity. All high volatile bituminous coals are suitable for sponge iron making
- Even lower rank coals give more reactive char (e.g., Lignite). But these are associated with two other problems:
    - Very high moisture
    - Tendency to become powdery and get air borne on heating.

After partial drying these can be used as thrown coal in sponge iron making.

- A fuel ratio value of lower than 1.5 (i.e., FC/VM < 1.5) ensures adequate reactivity

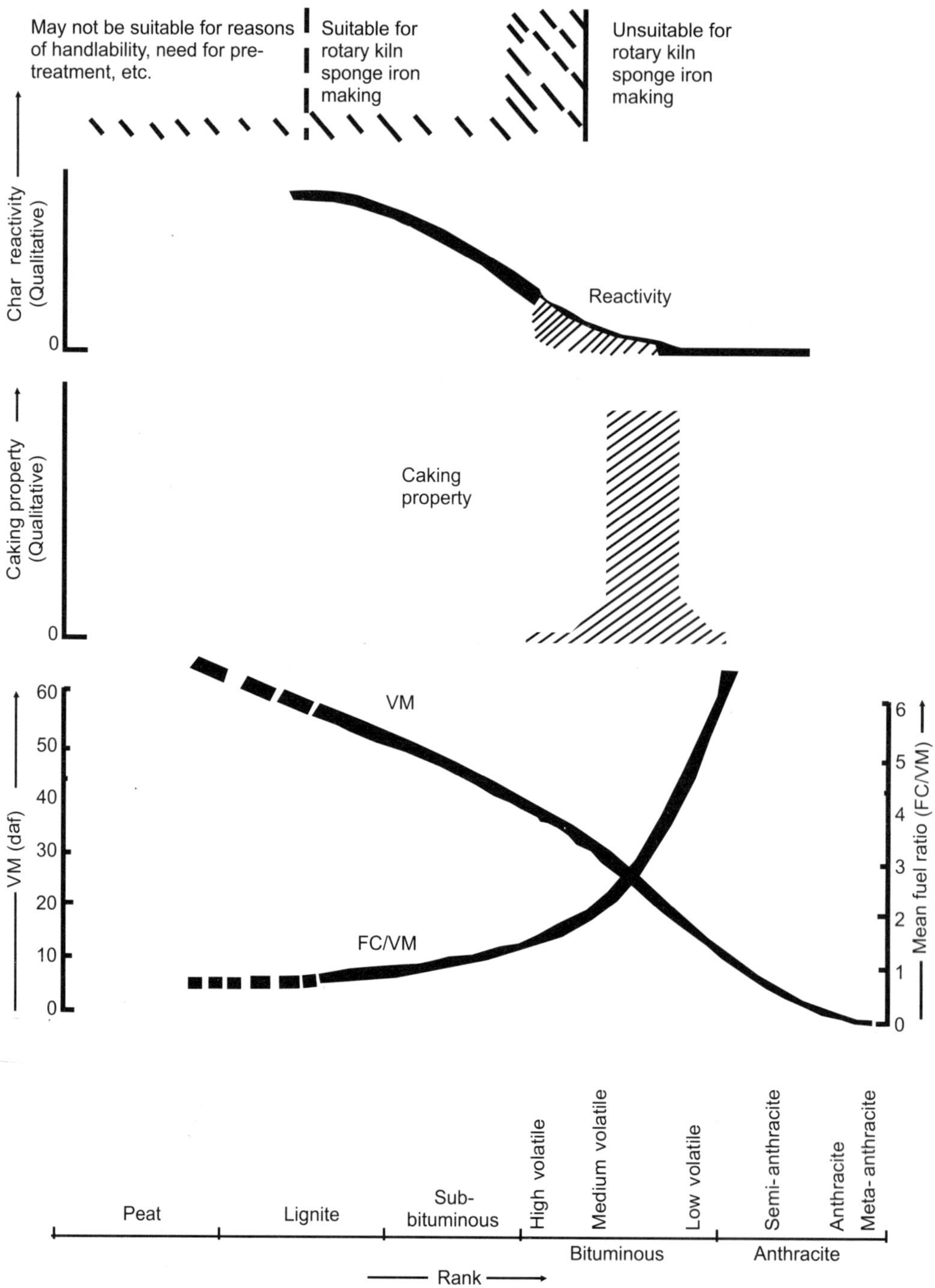

Fig. 9.7 Chart showing guidelines for coal selection

- Coals with fuel ratios up to 1.8 generally have acceptable reactivity
- Caking and swelling coals yield low reactive chars and are not suitable for sponge iron making.

  For the same reason, any type of coke is not suitable for sponge iron making.
- Pre-reacted, re-circulated and higher density chars have poor reactivities, since these chars have been exposed to high temperature for a long time.

  Figure 9.7 gives a qualitative diagram for selection of coal based on reactivity.

9.15 OTHER CRITERIA FOR COAL SELECTION

Ash amount and initial deformation temperature of ash under mildly reducing condition are the other criteria to be considered for coal selection. An increase in ash content by 1% reduces kiln productivity by 2%. But this figure is much less than the damage ash causes in other metallurgical reactors. Where ash needs to heated to much higher temperatures, melted, flux is provided for slagging of ash, flux needs to be heated and calcined for which additional coal/coke (with addition ash) has to be added. In contrast, in rotary kiln, ash only occupies kiln volume and needs to be heated to the processing temperature. Ash, thus, does least harm in sponge iron rotary kiln as compared to other metallurgical reactors.

A low fusing ash manifested by a low Initial Deformation Temperature, tends to form agglomerates when used in the thrown coal. It also increases tendency of accretion formation, although this aspect has not been conclusively proved. Using such coals as feed coal has been found not to give any problem in operation, as the char from feed coal rarely sees temperatures in excess of 1000°C.

Chapter 10

Accretion or Ring Formation Inside a Rotary Kiln

Accretion or Ring Formation is a perennial problem, which all the rotary kiln sponge iron makers have to face and live with. When a freshly lined kiln goes on production, the surface colour of refractory slowly darkens as it absorbs iron oxide. After some duration of operation, and especially if the process goes temporarily out of control, deposits are seen building up at various places on refractory surface. Deposits may be localised or, more commonly, may be forming around the circumference (like a ring, hence called ring formation). Sometimes the deposit formation is also accompanied with formation of agglomerates or partially fused association of solid particles. Agglomerates come out into the product discharge circuit but, sometimes get attached to the deposits on the refractory surface and help it to grow in size rapidly.

Ring formation makes the inner diameter of kiln progressively smaller at certain locations. While this renders uniform movement of charge extremely difficult, in extreme case of ring formation the kiln diameter may get fully blocked forcing a stoppage of the production campaign. During a shut down these 'Rings' or blockages have to be removed using pneumatic concrete breaking device, or any other suitable means.

10.1 CAN RING FORMATION BE AVOIDED?

While the problem cannot be eliminated, there are means of control of accretion and early detection and management of any large deposits, which might be forming on the refractory surface.

It is assumed that a lower ash fusion temperature of coal means a higher propensity to accretion formation, but this thumb rule has not been properly substantiated yet. Ash fusion temperature (the Initial Deformation Temperature or IDT) is almost always above 1150°C. The refractory surface of a properly operating kiln would never reach this temperature. Those ash compositions with IDT close to 1150°C are high in FeO, while those with IDT 1250°C and above are low in FeO. In rotary kiln condition where FeO is in plenty the above difference in FeO content would not matter.

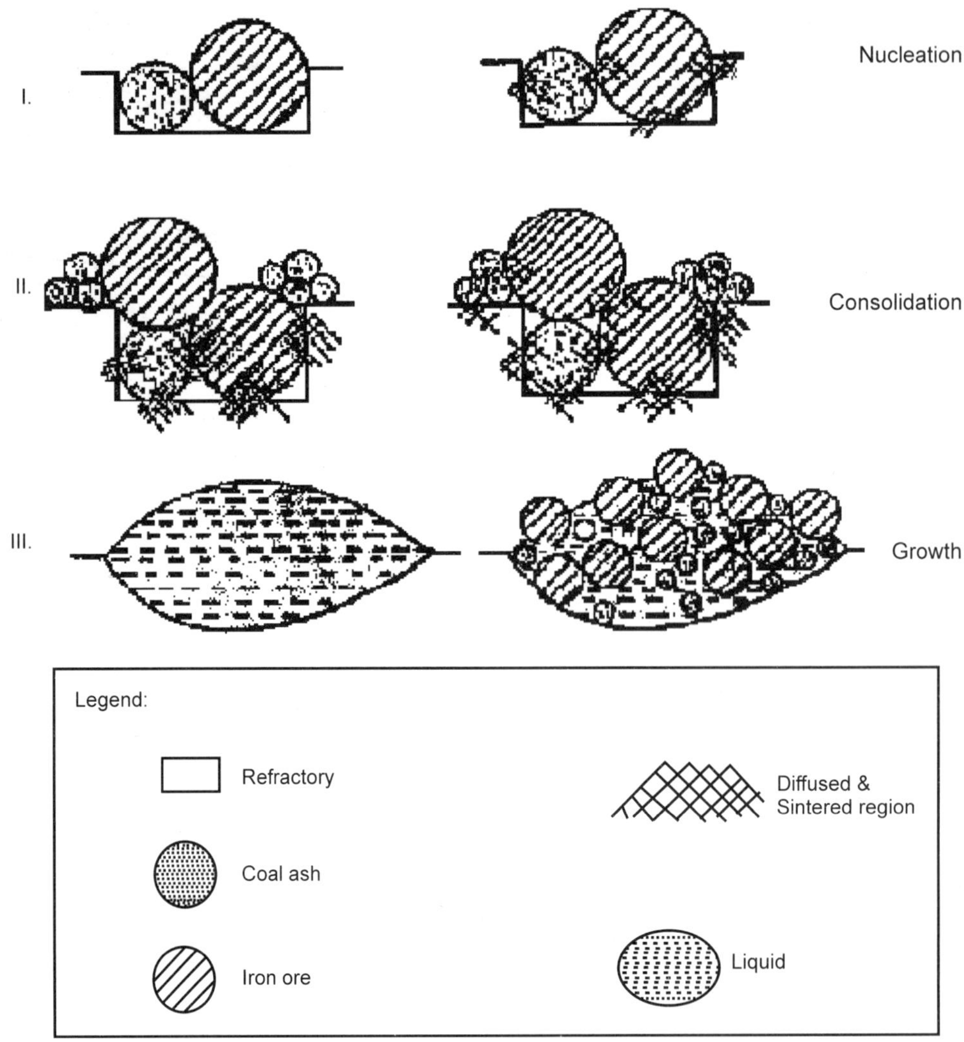

Fig. 10.1 Mechanism of initiation and growth of accretion

10.2 DEPOSIT FORMATION THROUGH SOLID STATE SINTERING

A particle of ash and of iron ore caught together in refractory crevices would slowly sinter over a length of time to reach a low melting composition (Fig. 10.1). But even the lowest melting composition has softening point of 1070°C and above (in the quaternary phase diagram FeO-SiO_2-Al_2O_3-CaO). Since discussing a quaternary phase diagram is rather complicated, let us ignore for a moment the minor constituent i.e., CaO (about 5%) in the quaternary. In the ternary phase diagram FeO-SiO_2-Al_2O_3, the lowest melting constituent melts at 1083°C (Fig. 10.2 & 10.3). We can make a simplifying assumption that accretion mainly comprises of $FeO + SiO_2 + Al_2O_3$ and the presence of flux may depress the fusion point by 13 to 15 °C.

Thus, ash and ore particles can sinter and are likely to make a composition with a fusion point of 1083°C, which can be lowered by 13 to 15 °C in presence of flux containing CaO and MgO, but even this composition would not normally melt in an efficiently controlled kiln. While deposits formed by solid state sintering are inevitable, it would not cause any hindrance to the production process as long as it remains solid. Only when it melts, it starts gathering other solids around it (mostly semi reduced sponge iron and ash) and re-solidifies into a larger mass. This generally happens when for some reason, the kiln temperature, which is generally maintained near 1000°C (actual temperature used depends on the characteristics of iron ore and coal) goes out of control. In majority of cases this happens because the thermocouple tips, which measure continuously the kiln temperature, get covered with deposits. Thus when the operator tried to control the kiln temperature to about 1000°C, the temperature of kiln is actually significantly higher than 1000°C.

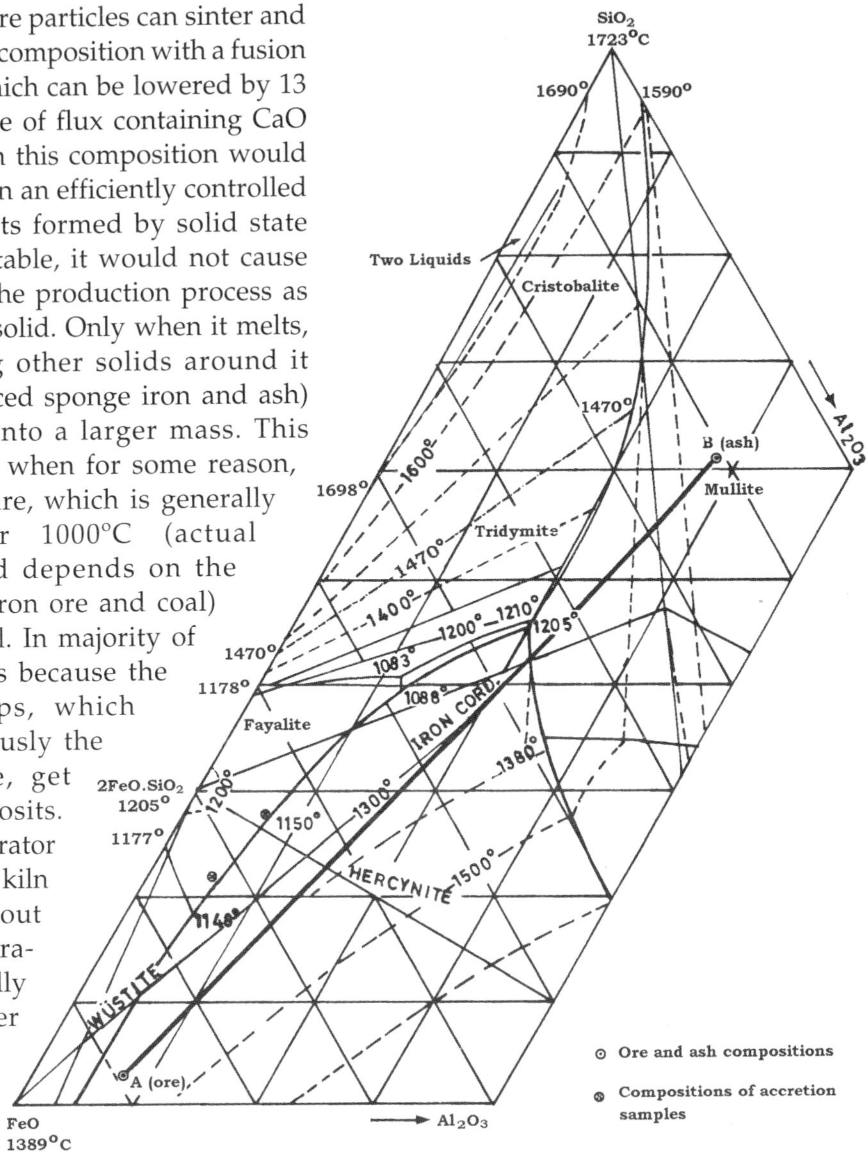

Fig. 10.2 FeO-SiO_2-Al_2O_3 phase diagram

10.3 REASONS FOR CATASTROPHIC GROWTH OF DEPOSITS

Since freeboard gas, and the refractory around, are always at higher temperatures than the solid material in any cross section of the kiln, a protrusion on a refractory surface receives higher heat radiation than rest of the refractory surface. The protrusion or deposit therefore acquires a higher temperature than rest of the refractory. Agglomerates, which are already partially fused, tend to get glued to the top of the deposit and form a bigger protrusion. It thus becomes a vicious circle. Once a protrusion is formed it becomes difficult to arrest its rapid growth.

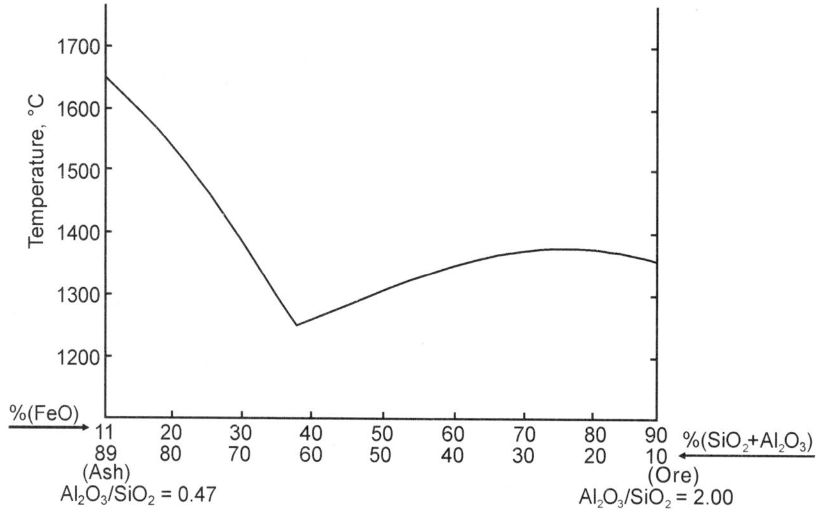

Fig. 10.3 Liquidus line in a typical ash-ore system

10.4 PRECAUTIONS REQUIRED

To prevent such happenings it is a general practice to measure the kiln temperature using quick response thermocouple (QRT which are made of thin mineral insulated thermocouples). These QRT's are inserted temporarily through ports on the kiln specially made for them or normally meant for collecting samples. A daily comparison of the records of QRT's and the corresponding permanent process control thermocouples helps a long way in circumventing the problem of Accretion Formation.

Kiln operators can use the QRT records in another way. When this thermocouple is inserted inside the running kiln, during rotation the thermocouple is for some time within the solid charge, while at other times it is sensing the free board gas. Consequently operators record the minimum temperature as the QRT-bed temperature and the highest temperature as the QRT-gas temperature. While the QRT-bed temperature is a relatively accurate estimate of the average bed temperature, QRT-gas temperature does not reflect the gas temperature at all. As the thermocouple tip 'sees' the refractory for more than 70% of time and the solid bed for the rest of the time, whatever heat radiation it receives from gas is being continuously radiated to the refractory and solid bed. Consequently, the recorded temperature is much less than the gas temperature. Since it 'sees' refractory surface for the longest time, the recorded temperature is closer to the temperature of refractory surface. Since the refractory surface temperature controls the fact that any molten particle falling on it either immediately solidifies and falls off or otherwise fuses to the surface and sticks and thereby starts formation of ring, control of this temperature is extremely important. The QRT-gas temperature at any point can, therefore, be taken to be a measure of "the accretion proneness" at that point.

10.5 ACCRETION REMOVAL DURING OPERATION

This is something which has been attempted since long but has met with very limited and uncertain success. Some accretions fall off on their own, particularly which are fragile. But, in sponge iron making we hardly come across such accretions. The accretions are usually quite hard and some of it are so adherent to refractory that on removal during shut down it takes away a layer of the refractory.

When accretion is in an accessible region, such as near the kiln mouth at discharge end, a long concrete breaker can be inserted to break it off. But very rarely accretions are so conveniently located. Gunning it down with regular bullets or heavier projectiles have been considered but has never been attempted to our knowledge as their use is highly dangerous both to the personnel and equipment. Same consideration applies to throwing of explosives on the accretion site.

What has been adopted with somewhat limited success is quenching the site of accretion with water during the operation itself. Although the results are not always positive, but the good thing is that no adverse effect of water quenching has been reported, which is rather surprising. Those who use this technique regularly say that there is no discernible adverse effect on the product quality. To date, this appears to be the only practicable method of removing accretion during operation even though the outcome is erratic. Some apparently have thrown common salt over the deposit site but have not reported positive result. Theoretically speaking, this measure would rather enhance the growth of deposit instead of controlling it.

Sponge Iron Properties: Re-oxidation and Auto-ignition of Sponge Iron

11

CHAPTER

11.1 REASONS FOR RE-OXIDATION TENDENCY

As mentioned earlier, the main rotary kiln product, the Sponge Iron or DRI is highly porous; the porosity could go as high as 52%. To this Figure of 52% we should add the porosity of the parent iron oxide. The porosity exposes a lot of surface area to the atmospheric air for oxidation and the sponge iron has to be cooled out of contact from any air, oxygen, moisture or even carbon dioxide. Once at room temperature the propensity to oxidise, also called pyrophoresity, is less acute but it still demands careful storage and transportation. Under normal atmospheric conditions, sponge iron re-oxidises and loses metallisation rather too slowly to be of major concern in a production facility, as long as wet conditions are avoided. Once exposed to rain, or even moist air, the loss of metallisation is rapid.

The rotary kiln DRI has got a bad name mostly because of its gas-based counterpart, which is highly pyrophoric and has caused many a fire with catastrophic effect. Shiploads have been burnt with massive damage to ships a number of times. Sponge iron burns without any flame. In a burning sponge iron heap one may see a short light blue flame at isolated locations, which is caused by burning of carbon monoxide, which is generated within the heap. We had once witnessed the spectacle of burning of shredded scrap bundles kept along with sponge iron, which had caught fire.

Coal based DRI is much more stable and no such accident has come to our knowledge so far. However, it does not mean that coal based DRI cannot catch fire. We had seen it catching fire when sparks from gas cutting torch were falling directly on it for a long time.

Modern commercial sponge iron are prepared under conditions which make it even more prone to re-oxidation, so much so that stocks and piles are now known to catch fire under apparently normal conditions of storage, usage and transport.

The phenomenon of auto-ignition of sponge iron is relatively newly discovered, even though the sponge iron has been known through ages and its susceptibility to oxidation is also well known since long.

11.2 HISTORY OF AUTO-IGNITION

Sponge Iron was first extracted by man about 4,500 years ago. This sponge iron was made in semi-fused condition and was immediately beaten to shape after re-heating. This immediate conversion practice was a ritual. This ritual was probably started because of the tendency of oxidation of sponge iron on exposure to atmosphere, which led to poor results, if the conversion to final tool was not immediate. Thus the practice of guarding it against re-oxidation was started even when the propensity to re-oxidation was not consciously realised.

With the advent of blast furnace, the practice of sponge iron making was relegated to insignificance. The primitive practice however, continued in many remote places including remote places of India.

During the World Wars, severe constraint was imposed on iron and steel resources, as the consumption in wars was very high. Germany started supplementing its iron needs during the wars using Krupp-Renn process which made sponge iron in a rotary kiln using coal. In this process the sponge iron was partially fused in the final stages of reduction to make what they called 'Luppen.' Since Luppen was fused and consolidated in the last stages it was not prone to re-oxidation. But the process was plagued with heavy deposit formation and fell into disuse when strain on iron resources eased after the World Wars. The Krupp-Renn process was to latter re-emerge as the Krupp-Codir process of sponge iron making. Till this time there was no recognition of the possibility of auto-ignition of sponge iron.

In 1950s Germany, Mexico and United States independently attempted to develop commercially viable sponge iron making processes, using reformed natural gas as reductant. Units based on HyL, Midrex, Armco and Purofer Processes came into commercial production, which made large quantities of gas based sponge iron available for shipment. From this point the problem of auto-ignition raised its head.

11.3 FIRST REPORTED INCIDENCE

A shipload of sponge iron was sent from Germany to Portugal by the sea route. The ship by name 'Cape Crest' arrived at Lisbon port in 1977 and the cargo was found to be on fire. It is not clear how much of the damage was caused to the ship and whether any part of the cargo could be saved. In May 1979 a shipload of sponge iron from Europe, after crossing Atlantic Ocean arrived at Houston, Texas in USA and the cargo once again was found to be on fire. In December of same year a sponge iron shipment from Canada was found to have caught fire on arrival at Barcelona, Spain.

There could have been other incidence of fire, which were not reported, but after these disasters stringent shipping norms were introduced. Ships for sponge iron were required to use McGregor Hatch (an airtight hatch). Apart from this, at the time of loading of sponge iron in a ship hold, the hold was required to be flushed with carbon dioxide gas. As it turned out this precaution did not give much reprieve.

In June 1981 two shiploads of sponge iron (probably from Indonesia) arrived at Vishakhapatnam port. During inspection both the cargoes were found to be OK. But during waiting period and by the time unloading could start, fire was detected in both the consignments.

To overcome the fire problem Midrex started hot briquetting the sponge iron into pillow shape. The resulting HBI was much more consolidated and stable for shipping purpose. HyL also followed suite. Midrex and HyL tried to produce as much HBI as possible but it was more costly than normal sponge iron. HBI was considered very safe and no fire hazard was envisaged. But there was a newspaper report that an HBI cargo on the way from Venezuela to Spain caught fire. This happened in March 1982. Midrex denied that it was HBI and claimed that it was a cargo of briquetted sponge iron from Fior process. It may be noted here that Venezuela operates all the 3 types of units: Midrex, HyL and Fior (now Finmet). Midrex stand was probably correct, as, at that time, the Venezuelan plant was briquetting only Fior sponge iron.

11.4 EYE WITNESS REPORT

We had only once the privilege of seeing a pile of sponge iron on fire. But there have been a number of incidences when our colleagues in the organisation had seen the incidences of fire.

In 1981 a shipment of sponge iron received from Indonesia was lying in stock in Alloy Steels Plant of Steel Authority of India Limited at Durgapur, in the eastern Indian state of West Bengal. When our colleagues had the opportunity to see the pile, the fire in the pile had been extinguished apparently by a very copious spray of water. The pile had turned black and was emitting evaporational steam. In the same year the sponge iron obtained apparently from the same source was lying in stock in the Bihar Alloy Steels Limited (BASL), an electric arc furnace based steel plant at Patratu in the eastern Indian state of Bihar (now Jharkhand). The pile had caught fire and was extinguished by heavy water spray before it could spread over a large area.

But in 1985 an identical incidence took place at BASL itself and this time the fire was a massive one and lasted over 3 days. Late in the second day we had reached site of fire. A part of the pile, which had taken a conical shape, because of angle of repose requirement, was looking normal with just a small bluish flame at the top of the cone. On approaching the pile we could feel the heat radiation from some distance itself. On removing a part of the apparently normal outer surface, a white-hot interior could be observed. The appearance from a distance was definitely misleading. The sponge iron pile on one side was overlapping with a pile of shredded scrap. The bales of shredded scrap in contact were also white hot, and the fire was spreading into the scrap pile as well.

With the help of a front-end loader we brought a small quantity of burning sponge iron into a relatively open flat space and we tried short experiments on this small burning pile. First we tried blowing nitrogen through the pile with the help of a perforated steel pipe. Then we tried carbon-di-oxide, followed by argon gas. All these only resulted in enhancing the fire, since the convection current associated with the gas flow helped to suck in surrounding air into the burning pile.

Since it was not possible to cover the large burning pile and keep it out of contact with air, such an experiment on the small burning pile was not tried. We tried spraying water. Soft spray did not have much effect. Evolving steam brought in air from surroundings just like the experiments with blowing in of inert gases. But heavier sprays were effective. It was therefore decided that with the help of front-end loaders and grab buckets, the unaffected sponge iron would be separated out from the burning portion and thereafter the burning part would be flooded with water.

This was done in the night and we could see that fire was extinguished successfully next day. Thereafter, we took a few samples and physically and chemically studied the burnt pieces. To our surprise even in the sample taken from regions, which were red hot, the particles showed only superficial oxidation. Substantial portion of the interior was unaffected and was in metallised form. This indicates that, if detected somewhat early, much of the value in burning sponge iron pile can be saved. Contrary the popular belief that water converts sponge iron back to Fe_2O_3, it does not happen to any significant extent during extinguishing of fire apparently because of the high heat evolution, which helps in removing surplus water by vaporisation.

11.5 INVESTIGATIONS RELATED TO AUTO-IGNITION

After the bitter incidences of auto-ignition described above, detailed investigations were taken up almost parallelly by M/s Lurgi Germany, HyL Mexico and Midrex USA. Details of the internal studies are not readily available but Mr. V.J. Clancey, UK subsequently studied, summarised and interpreted the findings to which we will return a little later.

Subsequently, National Science Foundation

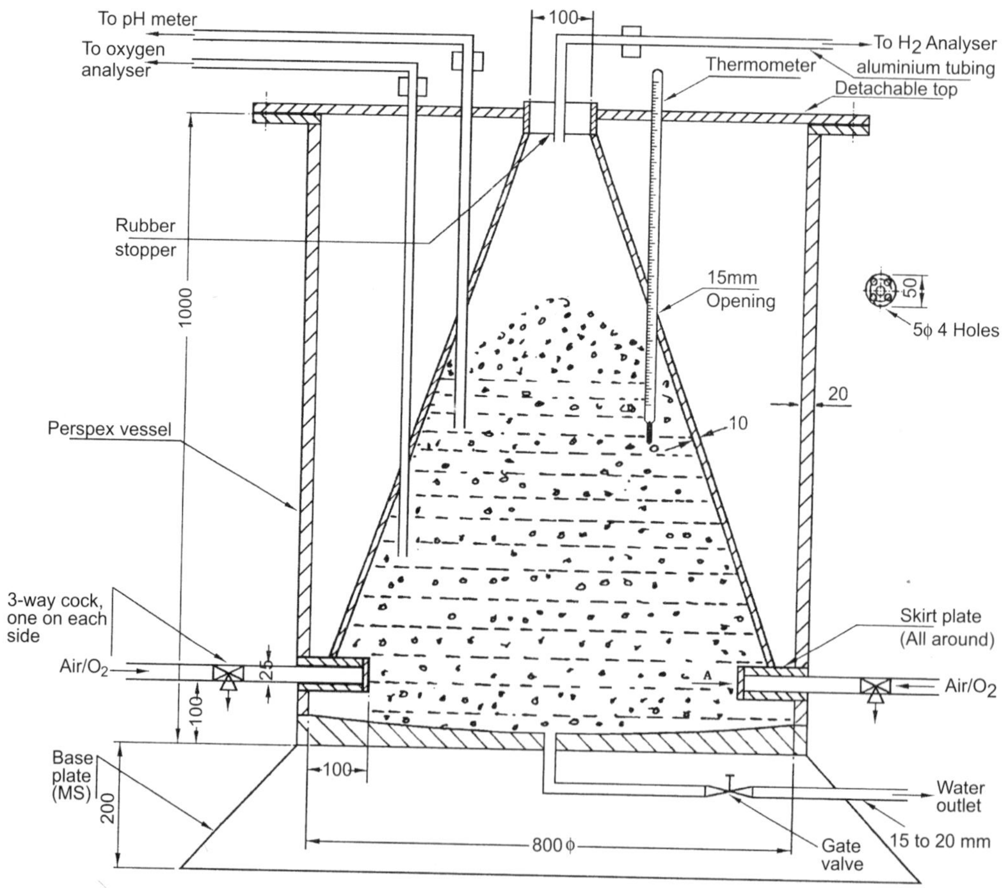

Fig. 11.1 Sketch of the experimental setup (Perspex Model) at RDCIS, SAIL

of USA sponsored a project to study the problem of Auto-ignition. Joint collaborators to this project were University of Pittsburgh, USA and the Research and Development Centre for Iron and Steel (RDCIS), Steel Authority of India Limited (SAIL), Ranchi, India.

In the investigations of RDCIS SAIL, which is the corporate R&D of SAIL, and is located at Ranchi, a Perspex model was made which simulated the hull of a ship. Coal based sponge iron samples were kept within this physical model and were made to react with water of different salinity in different prolonged experiments. Each experiment lasted for at least 3 to 4 days. The gases evolved could be collected within this model and these gas samples were analysed. The gases were found to be very high in hydrogen level and higher salinity of the water used resulted in higher volume of liberation of hydrogen.

The temperature rise during the different experiments was also measured. While temperature increase was observed and there was some trend of higher temperature rise with higher salinity, but the extent of temperature rise was marginal as well as inconsistent. These investigations did not lead to any significant insight towards the reasons for auto-ignition, but were otherwise useful in providing aqueous corrosion rate data of coal based sponge iron and in confirming the mechanism of aqueous corrosion of sponge iron.

RDCIS, SAIL further carried out EDAX investigation of the corrosion product obtained in the above experiments. The results supplied some missing links to Prof. Neil Birks at the University of Pittsburgh USA and helped him in formulating a plausible hypothesis on auto-ignition of sponge iron.

Fig. 11.2 Photograph of the perspex model

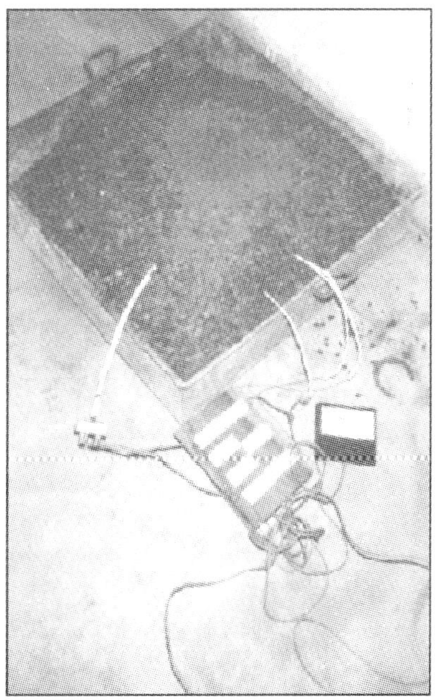

Fig. 11.3 Photograph of an experimental setup to study temperature rise during aqueous corrosion of sponge iron

11.6 OTHER SIGNIFICANT RESULTS

In an investigation carried out somewhere in Europe, a fairly large gas based sponge iron heap was kept in open atmosphere for nine months including a full three months of rainy season. After nine months it could be seen that the surface had turned reddish brown and surface samples showed almost zero metallisation. The surface had also become virtually impervious obviously because of increase in volume associated with corrosion.

On the other hand when samples were taken about 10 cm below the surface, the samples were black. The samples had lost very little of its metallisation and there was no evidence of substantial temperature rise inside the pile.

In another similar set of experiments, fresh gas based sponge iron was piled over an already corroded pile. After sometime, measurements inside the pile showed evidences of temperature increase. Although several variations of this were tried, and all pointed to increase in temperature, but in controlled experiments the increase in temperature was hardly of the magnitude, which could justify and explain auto-ignition.

11.7 CLANCEY'S HYPOTHESIS

As mentioned earlier Mr. V.J. Clancey of U.K. studied the actual occurrences of fire in cargoes of sponge iron and also the related reports made by Lurgi Germany, HyL Mexico and Midrex USA. He derived some important conclusions and presented his hypothesis on the reasons for enhanced oxidation of sponge iron during shipment but could not pinpoint the exact reason for triggering of auto-ignition.

Clancey observed that sponge iron is highly surface active material because of its high porosity (over 50%) and high surface area (3000 to 9000 cm^2/gm). This makes sponge iron capable of adsorbing on its surface a number of gases from atmosphere. Sponge iron has a stronger affinity for adsorbing moisture from atmosphere even in relatively dry weather. It is not uncommon for sponge iron under ambient condition to adsorb about one percent moisture. However, the amount adsorbed is highly dependent on humidity and temperature.

Higher temperatures tend to decrease the quantity of adsorbed moisture. When the sponge iron is charged in melting bath almost entire adsorbed moisture is released without loss of metallisation. Since this aspect does not cause any problem during melting, such users tend to completely ignore this aspect.

But things are different when sponge iron is transported in the hull of a ship. In a typical shipment there would be about 6000 tonnes of sponge iron, which would carry with it about 60 tonnes of adsorbed moisture. Due to temperature fluctuation during transportation as much as 30 tonnes of this adsorbed moisture may sublime and condense on the walls of ship's hull. After sometime this water may drip down and collect at the bottom. Thus a part of cargo may become wet even if there has not been any leakage in the ship's hull. Since it is very common for the walls of ship's hull to have salt deposits, lower layers of sponge iron would thus get wetted by saline water leading to enhanced corrosion.

Clancey thus pointed out reasons for enhanced danger during shipment of sponge iron but could not explain any further.

11.8 HYPOTHESIS OF NEIL BIRKS

Neil Birks of the University of Pittsburgh USA accepted Clancey's observation and

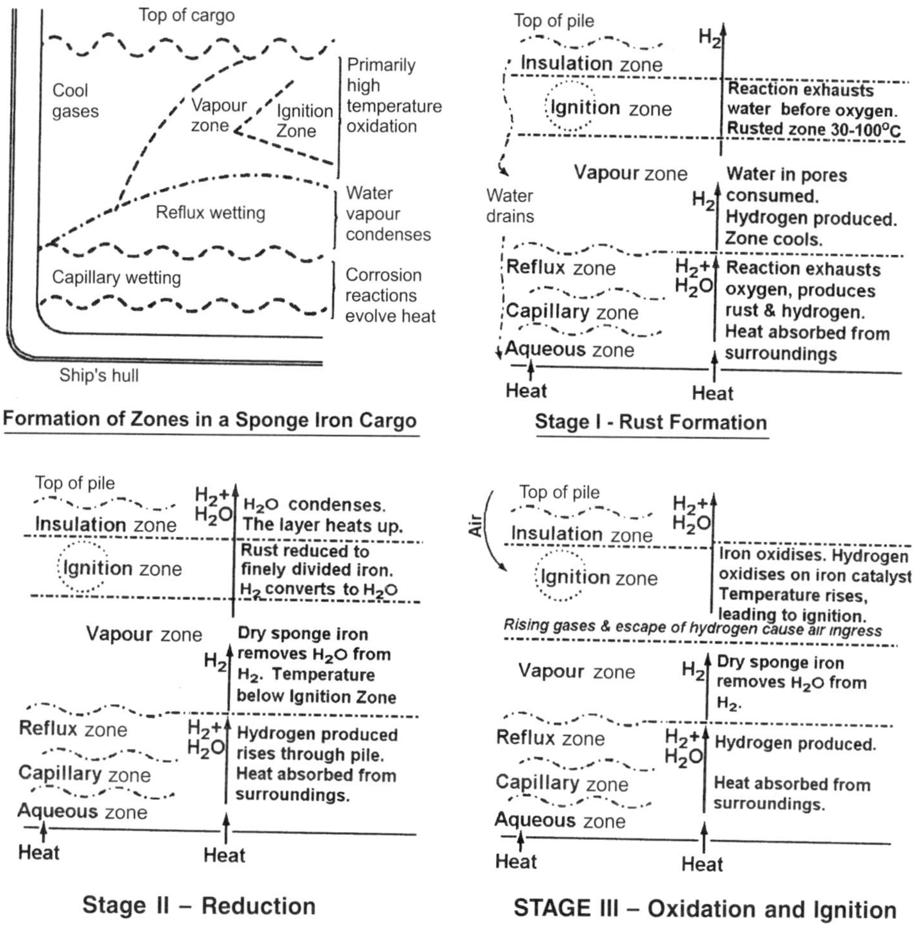

Fig. 11.4 Zones in a sponge iron cargo and stages leading to auto-ignition (Birks)

realised that the situation in the hull of a ship becomes similar to the cases where fresh sponge iron is piled over wetted and corroded sponge iron. He carried his reasoning from that point and also studied reports related to burnt cargoes. He realised that fine powdery and highly pyrophoric sponge iron was found in most cases of burnt cargo. Since such a sponge iron was never stacked to start with, this must have been generated in-situ. Prof. Birks carried out a few controlled experiments simulating the ship's hull and could find evidence of in-situ formation of such fine powdery sponge iron.

Prof. Birks thus evolved a hypothesis that auto-ignition of sponge iron takes place in the following manner:

(i) When sponge iron gets wetted by water it undergoes an electro-chemical corrosion and gets converted into oxides and hydroxides. In the process there is liberation of heat and also there is liberation of hydrogen.

$$Fe + (x+1)\ H_2O = FeO.xH_2O + H_2$$
$$\Delta H = -ve$$

(ii) Since sponge iron is highly porous it acts as a heat insulator in spite of its high metallic content. Therefore corrosion generally results in local over-heating.

(iii) In a ship's hull the lower layers of sponge iron are likely to get wetted by water. This wetting may be through leakage or other reasons as explained by Clancey. The H_2 gas thus liberated by wetting and corrosion may get entrapped in different portions of sponge iron pile.

(iv) Since corrosion also raises the temperature this may help in reduction of iron oxide available in the vicinity by the entrapped H_2. Since the reduction takes place at a relatively lower temperature the surface area of the sponge iron produced would be much higher than normal sponge iron and is therefore highly pyrophoric.

(v) Over a period of time such highly pyrophoric sponge iron may be generated at a number of places. Once there is a leakage of air into such an area, oxidation may start leading to elevation of temperature. It may happen that in a certain volume, oxygen leakage into the central portion takes place well after the surrounding areas are oxidised and thereby the temperature rises. This temperature may rise above the 'ignition point' of the freshly generated sponge iron at the centre and under these conditions, when air and oxygen leaks into this area, the sponge iron automatically ignites.

11.9 IGNITION TEMPERATURE OF SPONGE IRON: CROSSING POINT

It may be noted that what we call ignition point is that temperature at which under normal ambient conditions, the temperature rise in sponge iron become uncontrollable. A quick method of determining this temperature is to determine the "Crossing Point Temperature." This determination can be very easily made in the laboratory by keeping a sample of about 500 gm of say 3 to 10 mm sponge iron in a half litre Pyrex/ Corning glass beaker. A thermocouple is embedded in the middle portion of the sample in the beaker. The beaker containing the sample is placed in an air oven or a muffle furnace, which has already been preheated and stabilised at about 150°C. The temperature of the furnace or oven is slowly and steadily raised at a rate of 2 to 5°C/ minute. Both the temperatures, those of the furnace and of the sample, are monitored and plotted against time. To start with, the furnace temperature would be keeping ahead of the sample temperature. But after reaching a certain temperature level the sample temperature would be crossing and going beyond the furnace temperature. The temperature at this crossing point of the two curves is recorded as a crossing point temperature of sponge iron and can be viewed as the ignition temperature of sponge iron.

The crossing point temperature of commercial sponge iron normally lies between 200 to 350°C. Coal based sponge iron, made in rotary kiln, has a crossing point temperature close to the higher limit. The crossing point temperature of gas based sponge iron lie in the middle to lower regions of this range. The sponge iron primarily reduced by H_2, as in the HyL process, has the lowest crossing point temperature among the commercial sponge irons.

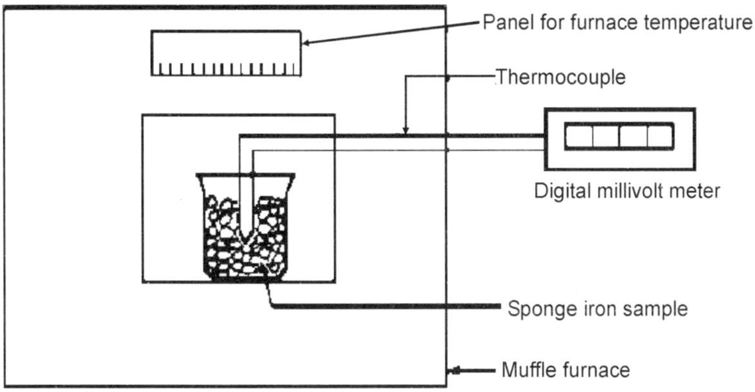

Fig. 11.5 Set up for determination of crossing point temperature

H_2 has got a very high diffusivity and small molecular volume. At relatively lower temperatures the reaction kinetics may become unfavourable for the reduction of FeO to Fe by CO; but reduction by H_2 would still be possible. Sponge iron produced by H_2 at lower temperatures in say, controlled conditions of a laboratory would have smaller pores in much larger numbers; and would therefore have much higher surface area than that of the sponge irons which we are used to. The crossing point of such a sponge iron may be even below 200°C. Formation of such a sponge iron somewhere inside a large pile would naturally make it very vulnerable to auto-ignition.

11.10 HOW VALID IS THIS HYPOTHESIS?

Confidence in the hypothesis of Prof. Birks would have increased if it were validated in control experiments. Unfortunately simulation condition of auto-ignition is extremely difficult and certainly not possible to reproduce in laboratory. It requires very large stack of sponge iron where a number of pockets of accumulation of H_2 could form.

One may argue that there have been reports of auto-ignition in the stockyard of a few units using sponge iron where there was certainly no wetting of the pile with water. But such reports of 'auto-ignition' may be viewed with suspicion. In a sponge iron using unit there are lots of source of ignition supporting objects (flying sparks, even scrapped artillery shells). Therefore these reports are unlikely to be true incidences of auto-ignition. However, these incidences do underline the fact that sponge iron is highly pyrophoric, hazardous and needs to be handled with care.

There has so far been no report of auto-ignition of sponge iron made by coal based route in rotary kiln. A study of their crossing point indicates that they have a higher ignition temperature. Since temperatures used in rotary kiln are somewhat higher than that used in sponge iron shaft, they (coal based sponge iron) have gone through the step of surface sintering, which reduces number of pores at the surface of the sponge iron. But question arises whether coal based sponge iron can be prone to auto-ignition even though the crossing point temperature is higher? If hypothesis of Prof. Birks were valid, even this sponge iron would be amenable to auto-ignition; only it should take somewhat longer time. Whether or not this happens can be confirmed only after large shipments of coal based sponge iron become commonplace.

11.11 PRECAUTIONS NEEDED IN SPONGE IRON STORAGE

In light of the above, we can identify a few essential precautionary steps related to handling and storage of sponge iron. Since the large surface area makes it prone to atmospheric re-oxidation, which is enhanced in presence of moisture, sponge iron should be handled and stored in such a way that it is kept away from water or moisture as far as possible. Therefore it should be kept in covered bay with hard floor having such an elevation so that water does not collect at the bottom.

But when large tonnages are handled, it is often not possible to find covered yard for all the sponge iron stocks. If forced to store in open, the pile should be kept undisturbed until use. Whenever reclaiming is started the pile should be quickly consumed.

Some 'Don'ts' should be closely followed. One should never store fresh sponge iron over a corroded pile. Such a process would contribute heavily towards proneness to auto-ignition. As and when fire actually starts in a sponge iron pile the only sure way to extinguish it is to use copious water spray or flooding with water.

Apprehension exists about the use of water. Like spraying water on calcium carbide only enhances the fire, some people think that water spray enhances sponge iron fire. While that is not correct, it should be kept in mind that light water spray, as also blowing in nitrogen or argon helps in enhancing convection currents which results in more air being made available to sponge iron and thereby the existing fire is enhanced.

11.12 CONCLUSIONS

1. Auto-ignition of sponge iron is the outcome of electrochemical corrosion of sponge iron. Moisture required for corrosion need not come from external sources and may remain loosely bonded to the surface of sponge iron. However, moistening of the surface alone does not lead to auto-ignition.
2. High porosity of sponge iron makes it highly surface active as well as a heat insulator.
3. Corrosion within the pile of sponge iron leads also to loss of permeability owing to volume expansion. Hydrogen gas liberated as a result of corrosion may get entrapped in some locations. This may reduce iron oxide available in the vicinity and thus may result in the formation of fresh and finely divided metallic iron of low ignition temperature. On gaining access to air auto ignition may start.
4. In control experiments auto ignition has not been observed in ambient conditions (only beyond 220°C).
5. A plausible hypothesis of auto ignition has been evolved, but needs to be validated.

CHAPTER 12

Uses of Sponge Iron

The production of sponge iron and use of sponge iron in steel making are both growing in India by leaps and bounds. The yearly production of sponge iron has already crossed 16 million tonnes. The coal based rotary kiln and the gas based routes are contributing about 60:40 in this production.

The demand for any item is directly dependent on its usefulness. And the market demand or success in the market is solely dependent on the faith of the consumers on its potential use. This was something lacking when tonnage sponge iron first appeared in the Indian markets in the 1970's (through imports) and the 80's (through indigenous production).

At that time, sponge iron was projected as a substitute for iron and steel scrap. As metallic content of sponge iron was much lower than scrap, sponge iron played second fiddle to scrap for a long time. But with the experience gained in its usage, many favourable properties of sponge iron were discovered, and sponge iron then turned out to be, at least in India, the second major metallic feed for steelmaking. It may possibly overtake the blast furnace hot metal in this regard is not so distant future.

The major reason for the wide acceptance of sponge iron is its closely sized granular nature. The close size range makes it amenable for continuous charging into reaction vessels. Continuous charging means that processing and charging can go parallel and there is no separate charging time. Further, during continuous charging in electric arc furnace (EAF), heat losses due to removal of roof for charging through bucket is minimized and sometimes completely eliminated. The sponge iron particles fall through refining slag and in the process melts and gets refined, further saving on refining time.

Other favourable aspects of sponge iron are its low sulphur content (generally around 0.02%) and absence of tramp elements (Cu, Sn, As, Sb, Pb, Cr, etc.). Significant levels of dephosphorisation have been reported while charging sponge iron, possibly due to the presence of oxides like FeO in the sponge iron. Further, FeO in sponge iron induces "boil" in metallic melt necessary for effective heat transfer and for aiding in the refining process. Separate boil inducing constituents are not

Table 12.1 Comparison of steel making in arc and induction furnaces

Item no.	*Criteria*	*Electric arc furnace (EAF) steel making*	*Electric induction furnace (IF) steel making*	*Remarks*
1.	Refining Operations	Possible	Not Possible	Removal of P, S & C is possible only in EAF
2.	Addition of Minor Constituents	Possible	Possible	Due to stirring by eddy currens, homogenisation of additives is more easily possible in IF
3.	Specific Power Consumption	Lower (increasese at lower capacities from about 450 to 700 kwh/tonne)	Higher (about 750 kwh/tonne fr 70% DRI charge)	For mini steel plants, the difference is small
4.	Electrical Fluctuations	High (higher peak demand)	Very Low (lower peak demand)	High electrical noise in EAF operation also leads to high acoustical noise. IF operation is relatively very quiet
5.	Tap to Tap Time	Larger	Smaller	Ultra High power (UHP) EAF's can have tap to tap time close to IF's
6.	Heating Configuration	Localised heating near the arc	All round heating in the meallic portion	Slag gets heated only by contact with metal in IF
7.	Heat Size	Higher (25 to 80/100 tonnes)	Lower (up to 30 tonnes)	
8.	Melting Rate	Lower	High	UHP EAF's have melting rates comparable to IF's
9.	Specific Refractory Cost	Lower	About 50% higher	
10.	Specific Capital Investment	Higher (complicated) power supply unit)	Lower (power supply unit is simpler)	EAF starts operation with short circuit condition. Current fluctuations are high; also heavy impedance is required in the circuit
11.	Skill Level Required	Higher	Lower	
12.	Versatility	Higher	Lower	Possible to make much larger number of grades in EAF's

required. However, some additional carburiser is needed.

12.1 USE OF SPONGE IRON IN ELECTRIC STEEL MAKING

Due to problems associated with gas based sponge iron (auto-ignition tendency) and relatively high cost of hot briquetting the DRI, the major portion of gas based sponge iron is consumed in-house by the producers either in their UHP-EAF or CONARC furnaces.

M/s Sunflag at Bhandara, Maharashtra are also consuming their sponge iron in-house although they are a coal based sponge iron producing unit. Many other coal based sponge iron producing units have now integrated steel making units in their operation. But most of them use induction furnaces for melting sponge iron.

The small merchant producers of sponge iron are mainly feeding induction furnace steel producers. Normally these steel producers are using 70% sponge iron in the charge with 10% cast iron and 20% steel scrap. The liquid steel produced which is about 70% of the total metallic charge, is cast either in pencil ingots or are continuously cast into billets.

In the 1970's and 80's most of the scrap in India were melted in electric arc furnaces (EAF's) and in a smaller number of induction furnaces (IF's). While use of Sponge iron in EAF's were attempted immediately, IF operators were wary of its usage since its induced a relatively violent reaction and necessitated changes in operation practice as well as design of operating platform.

But, with the experience gained, charging of sponge iron in induction furnace has really picked up. Even though electric power consumption is higher per tonne steel, at small capacity levels the difference is hardly significant, while electrode consumption is totally absent. Since there is no arcing there are no sudden power fluctuations (i.e., there is absence of electrical 'noise').

In Table 12.1 a comparison has been presented of the steelmaking in EAF and IF using sponge iron. No wonder, mini steel producers using sponge iron prefer adopting induction furnace route and it would remain their preference as long as the market demand is consistent with the inputs. Needless to say, there is very limited room to diversify with such an installation.

In Table 12.1 low skill level requirement is mentioned for IF operation. But skilled removal of the slag (or rather the oxide scum) is required all the time to allow sponge iron charge to meet the liquid metal without hindrance. Sponge iron would ignite and burn with atmospheric oxygen if it is held even momentarily above the surface of the melt.

Sponge iron requires higher energy for melting as compared to scrap. But by increasing the rate of heat input along with continuous charging, this drawback has been largely overcome. Being granular, it starts melting from all sides (i.e., all faces of a sponge iron particle starts melting simultaneously) and "melts like butter" and this is not possible with heavy melting scrap. Faster melting ensures lesser radiation loss and therefore, the higher energy requirement is largely compensated.

12.2 USE OF SPONGE IRON IN SUBMERGED ARC FURNACE

In a Submerged Arc Furnace (SAF), the electrical arc struck between the two electrodes remains covered with the solid charge material. Consequently this furnace is more energy efficient as compared to EAF. But since there is no access available for operators to reach and work the melt, this unit is not conducive to quality steel making. Instead the melt can easily be allowed to get saturated with carbon and then we get hot metal.

This route of making hot metal cannot presently compete with the blast furnace iron making process. But there is a possibility of making quite low sulphur and low phosphorous hot metal which would be suitable for conversion to ductile iron or spheroidal graphite cast iron. In 1990-91 the National Metallurgical Laboratory (NML) carried out extensive trials in its pilot 500 KVA submerged arc furnace, with the assistance of SAIL. The trials proved the versatile nature of SAF in that very inferior DRI could also be melted without any problem. During the trials, hot metal of quality suitable for conversion to ductile iron was always obtained.

Although this concept is being used in New Zealand and probably also in Myanmar and South Africa for making hot metal, but these are examples of commercialization with a special type of raw material or for specific site conditions. We cannot say that, at the moment, the practice of melting of sponge iron in SAF to make hot metal has been commercially adopted.

12.3 USE OF SPONGE IRON IN IRON MAKING

Since sponge iron composition is closer to steel than pig iron, its use as raw material to making pig iron in blast furnace appears at first, a wasteful exercise. It is indeed not favoured at present since the input cost of sponge iron does not make it economically viable for pig iron making.

But at times of hot metal shortage, the overall plant economics often justifies use of sponge iron in enhancing production from blast furnaces as well as save on expensive coke usage. Some technical disadvantages of sponge iron in blast furnaces have been pointed out, but these can all be overcome by obtaining from the producers sponge iron of slightly higher size (say + 8 mm) while compromising slightly on the metallic Fe content (say 77% as against the usual 80%). The other disadvantages of higher CO in the top gas and higher gas temperature would hardly be observable at the charging levels being contemplated (i.e., 10% metallisation of burden, approximating to about 9% of iron oxide feed).

The effect of charging sponge iron in small blast furnaces have been studied extensively in the past by U.S. Bureau of Mines, U.S. Steel Corporation, Steel Company of Canada and Ahmsa (Altos Hornos de Mexico S.A.). The results are summarised in Fig. 12.1. This plot indicates that, for 10% metallisation of burden, both productivity and coke rate improve by about 8%. But at higher rates of charging the benefits change. Decrease in coke rate drops to 5.5%. Productivity improvement to 9.5% is indicated, but this increase may not be statistically significant. The recent experiences of Tata Steel and Steel Authority of India Limited have been on similar lines. Because of decreasing benefits, charging of sponge iron by more than 10% may not be contemplated anytime in future. But even up to 10% the technical benefits are immense. It is estimated that sponge iron charging would be economically viable if the price of sponge iron is up to 40% of the price of coke. There was a period in the year 2005 when Durgapur Steel Plant experienced coke shortage and was purchasing additional coke at about Rs. 22,000/- per tonne. Sponge iron was available at Rs. 9000/- per tonne and the coke shortage was mitigated by the use of a small percentage (about 5%) of sponge iron. The situation has now eased, but relative to sponge iron the average price of coke has to go up only.

The Middletown plant of M/s A.K. Steel in the United States has been operating their blast furnaces consistently at very high productivities since 1996. Productivity of 4.37 t/m^3-day was reported in the year 2004 but the measures taken were not explained. They were known to be charging HBI (hot briquetted iron, i.e. compacted sponge iron) and high rates of the same appear to be the major reason behind increased productivity.

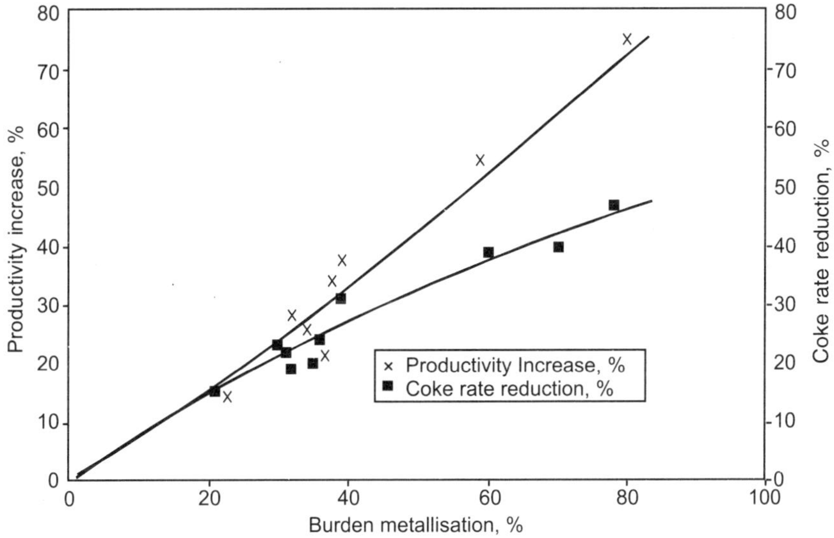

Fig. 12.1 Effect of burden metallisation on productivity and coke rate in blast furnace

12.4 USE OF SPONGE IRON IN CUPOLA

During 1980's the Kulti unit of India Iron and Steel Co. Ltd., were operating cupolas, in one of which sponge iron was charged to the extent of 20% of metallic charge. No problem was observed during sponge iron usage. Even though increased resistance to blast and higher coke rate were anticipated, these were found negligible. Flux requirement increased and it was felt that the sponge iron probably travels faster down the shaft as it is smaller in size. Extra coke charging before start of charging sponge iron helped in smooth operations.

If the above precautions are taken there should not be any problem in use of sponge iron as cupola feed. Only lack of economic viability is preventing use of sponge iron in cupolas.

12.5 USE OF SPONGE IRON IN OXYGEN STEEL MAKING

The maximum amount of steel is presently produced through hot metal and basic oxygen furnace (BOF) route. BOF steelmaking requires coolant during oxygen lancing so that the finish temperature does not go higher than desired. The best coolant for BOF is the revert steel scrap which simultaneously increases the yield by 18 to 30%. Sponge iron can very well be used to replace steel scrap. With wide adoption of continuous casting, quantity of revert scrap is going on decreasing, and presently we see sponge iron as the only alternative. Cooling capacity of sponge iron being marginally higher, it is required in lesser quantities so that the final yield of steel is 2 to 3% lower. Sponge iron carries with it a small quantity of FeO which may increase the tendency of slopping of

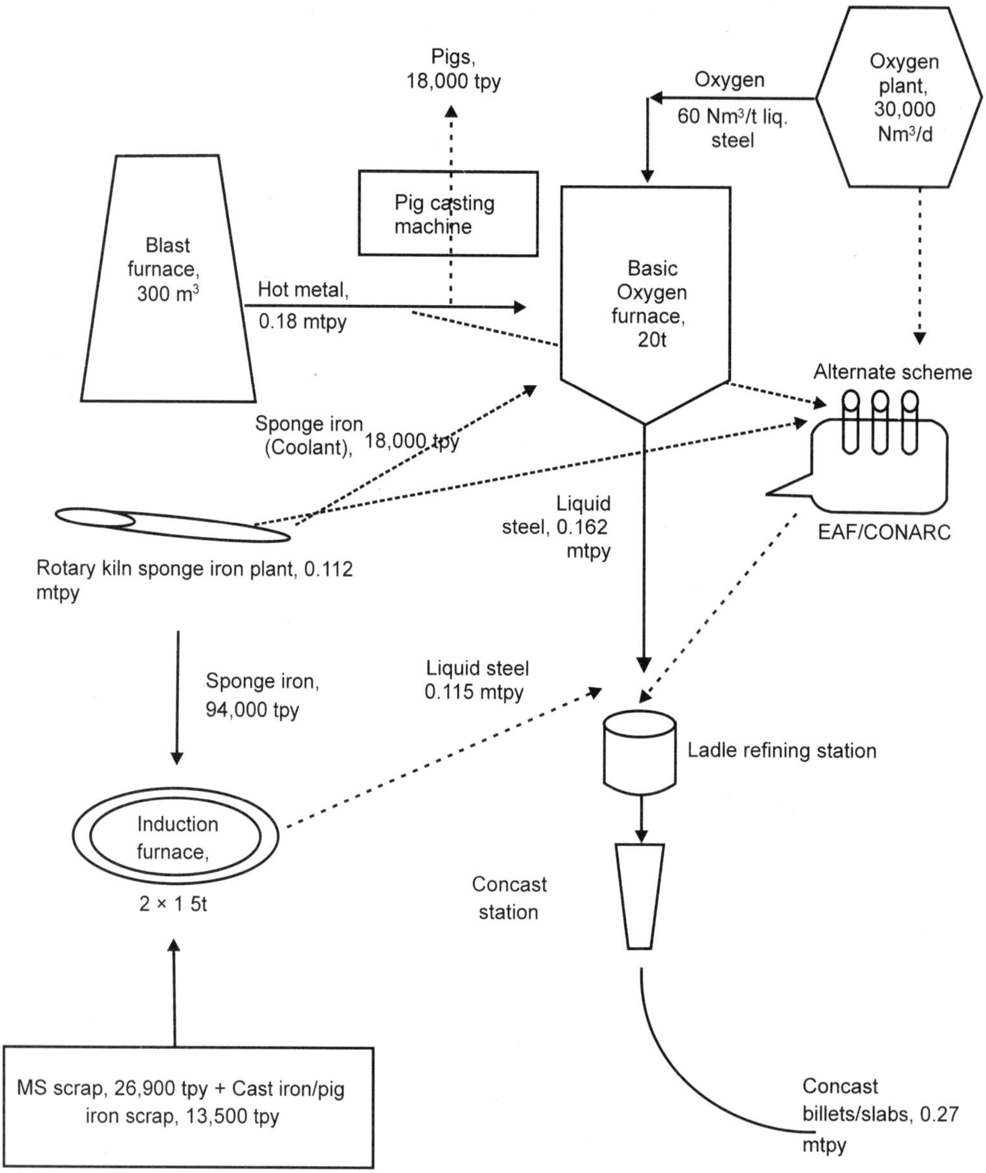

Fig. 12.2 Schematic of a typical mini/midi steel plant of India

BOF, but this aspect has not been conclusively established in the trials conducted in India. Sponge iron is poised to become a standard coolant in BOF operation.

12.6 SPONGE IRON AND HOT METAL: PARTNERS IN PROGRESS

Most of the entrepreneurs contemplating newer steel making units have realised the importance of marrying together the use of blast furnace hot metal and sponge iron in steel making stage. This is done either in an oxygen converter where sponge iron is relatively a minor input or in an electric arc or induction unit where hot metal would be a minor input among the two.

By using them together, the disadvantages of higher than desirable carbon in hot metal and undesirable iron oxide in sponge iron are compensated by each other. Further, during use in electrical furnaces, the heat liberated in the oxidation of carbon helps in saving a part of the costly electrical energy. It is little wonder that most of the sponge iron producers are contemplating expansion with a mini blast furnace and vice versa (i.e. mini blast furnace based units plan to go for sponge iron production). It is expected that mini-and midi-steel plants of India may follow schemes depicted in Fig. 12.2.

Process Design, Engineering and Operational Aspects of an RK-DR Plant

13

CHAPTER

When an entrepreneur thinks of utilising the natural resources of iron ore and non-coking coal deposits to make sponge iron to meet the market demand of the region, which we are presuming he has already assessed, then he looks for a process supplier who can give a suitable technology. Technology suppliers in sponge iron area have now proliferated. Even the technologies, which started with widely different concepts, have now converged to a nearly common process flow typically indicated in Fig. 2.2. Fig. 13.1 represents a typical raw material preparation, handling and proportioning unit while the product handling system is represented in Fig. 13.2.

The technology supplier develops his know-how based on past information, experience and scientific principles and interpolations. Once his technology has proved successful he has only to update his old document to cater to the next customer. But each customer has his unique need. The process of converting the customer's need into a project document is referred to as Process Design and Engineering – incorporating two overlapping steps of Process Design (also referred to as Basic Design) and Process Engineering.

13.1 PROCESS DESIGN AND ENGINEERING

Process Design refers to the clear understanding and determination of all process control variables and their optimisation and actual design of the equipment and facilities necessary for carrying out the process.

Process design aspect has been mentioned in chapter 6 (section 6.2) where the volume requirement of a particular capacity kiln has been discussed, while discussing the importance of determination of residence time in a rotary kiln.

When an entrepreneur gives his requirement to the technology or process supplier, he gives the following information:

1. The size of plant required;
2. Sources of Raw Materials, Utilities and Power
3. The plant site.

The process supplier arranges to test the raw materials for determining the suitability for the process, the productivity expected and

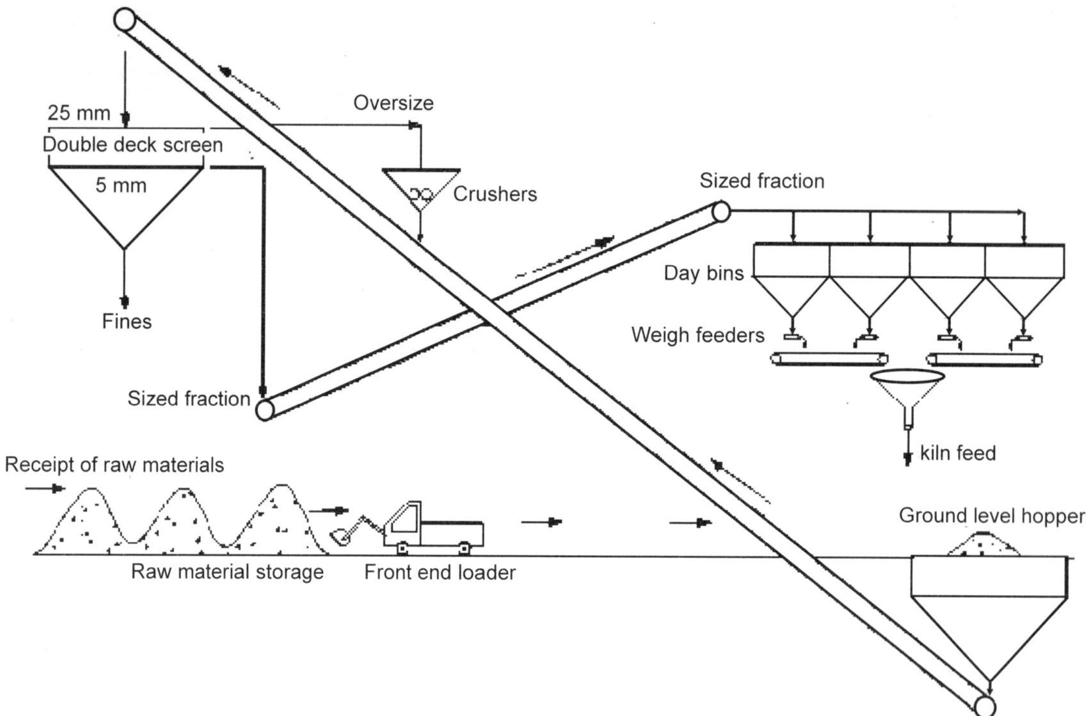

Fig. 13.1 Schematic of raw material handling and proportioning unit

process consumptions based on the properties of the raw materials. For interpreting test results and other input information to project the data on productivity, consumption pattern, operating parameters, etc. the process supplier needs to have nomograms or software or model of the process. He also needs to detail out processing techniques and prepare an operation manual. It is also required that he draws the overall schematic of the process, prepares specifications of main technological units and of critical equipment and components and prepares basic engineering drawings thereof. In short, he is now doing the process engineering, and this also encompasses the steps of project definition, preparation of feasibility report and project engineering and implementation which follow. Table 13.1 lists the steps involved in process design and engineering.

13.1.1 Raw Material Testing

For a DR plant, normally a Rotary Tube (Salvis) furnace is used as a bench-scale testing device. This test can predict the desirable process parameters, viz., residence time, temperature, degree of fill, etc., needed for obtaining good metallisation.

While ore reducibility and coal char reactivity can be measured individually (please refer chapter 9) and based on the results obtained a ruling can be given that a particular combination would be suitable for sponge iron making, but confidence in test result would certainly be better if tests are carried out in conditions closely simulating the actual conditions in rotary kiln. One such test is the Rotary Tube test, very commonly referred to as Salvis test (IS:11284-1985).

In fact, this test is relatively simpler than

both the Ore Reducibility and Coal Char Reactivity tests. So usually, the ore-coal combination is first tested in Rotary Tube test (after chemical and proximate analysis). If the combination does not pass the test then only ore and coal are individually characterised to decide which one is the real, or more of, the culprit. Apart from the compatibility test, the Rotary Tube also gives indication of the agglomeration tendency of coal (caking propensity) as also the decrepitation tendency of iron ore (tendency to generate fines). Both these are very important indicators for rotary kiln operation. An agglomerating coal is bound to give a char with low reactivity, while decrepitated ore is not only loss to the process but it further increases the propensity to accretion or ring formation.

Figure 13.3 gives a diagram of the equipment for Rotary Tube Test. It consists essentially of an electrically heated tubular reactor of 150 mm internal diameter and 650 mm reaction zone length. The tubular reactor is rotated at a predetermined rpm while nitrogen gas is passed continuously throughout the experiment. After the tube is heated to a predetermined temperature (generally 400°C, 700°C or 1000°C), iron ore (5 to 20 mm) and coal (1 to 15 mm) in predetermined ratio is mixed and charged into the hot zone of the tube. The temperature of tube is then raised to the maximum

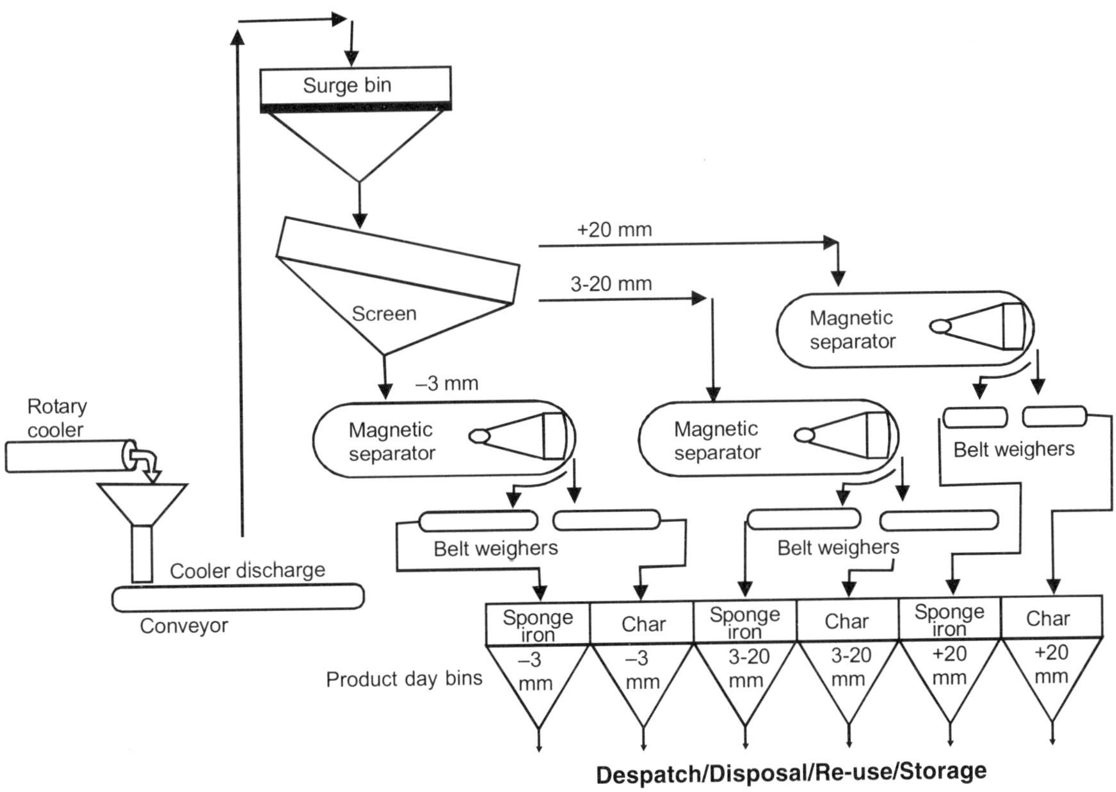

Fig. 13.2 Schematic of a product handling system

temperature (1000 to 1020°C) and maintained there for one and a half to three hours. Thereafter, the mixture is cooled within the Rotary Tube furnace under continued flow of nitrogen to room temperature. The product is magnetically separated and both fractions are size analysed and thereafter chemically analysed. The percentage of +15 mm of char is a measure of agglomeration property of coal while –3 and –1mm of magnetics give an indication of the decrepitating tendency of iron ore or its friable nature. Friable iron ores are not suitable for sponge iron making in rotary kiln without pelletisation.

The overall magnetisation (Fe(M)/Fe(T)) of the magnetic fraction should be above 90% for the combination to be commercially viable.

A successive phase, i.e., Pilot Plant testing, may then be taken up to generate data for the ultimate full scale plant and also produce small quantities of sponge iron for research and trial purposes. As the flexibilities in a Pilot Plant are more, the effects of changing design or operating conditions can be exhibited more quickly and economically than that would be possible on a full sized prototype.

13.1.2 Upscaling

Let us look at the Process Design problem from the process or technology developer point of view that ultimately has to supply the Process Technology either directly to the entrepreneur or through an intermediate technology supplier. Normally his design of the process is based on experience of operation of similar sized plants and units. But sometimes, the requirement of the entrepreneur is of a different size plant, in which case the process developer has to resort to upscaling (or downscaling as the case may be). Help is then taken from the mathematical model of the process (Chp. 5), which has been prepared based on data generated from earlier experiments and operating experience of Pilot Plants and other units. The model is validated and modified, if necessary, using data of some operating plants. Thereafter the mathematical model of

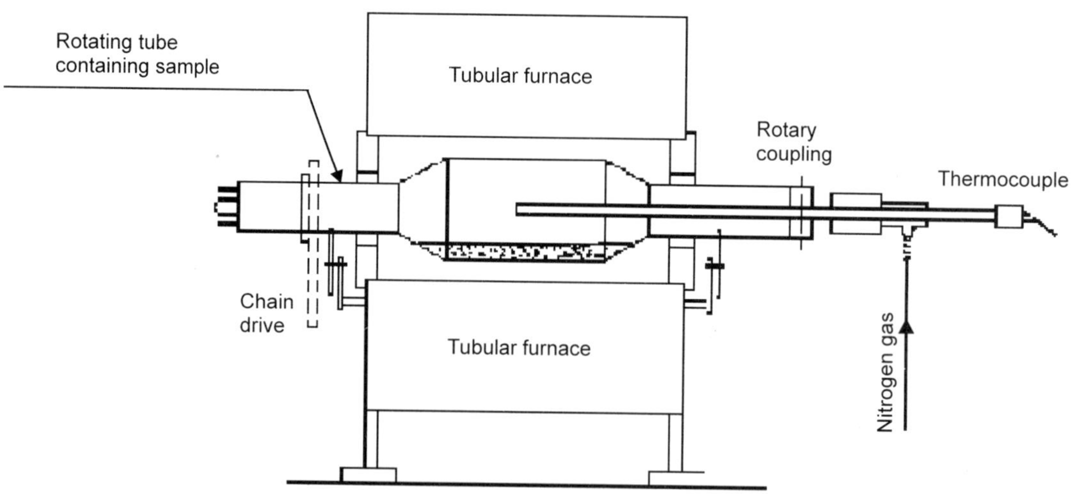

Fig. 13.3 Laboratory rotary tube furnace

process is proportionately up-scaled. Corresponding process flow diagrams are then prepared and then the individual equipment are designed.

We have talked about proportionate up-scaling. This 'proportionate up-scaling' may not always be linear and the designer has to draw upon his experience, and the experience of others in the field, to decide upon the method of this 'proportionate up-scaling.' Sometimes a subjective decision would have to be taken.

13.1.3 Sizing of Key Equipment

Based on the data generated and the size of plant required, the reactor volume can be worked out based on the example in chapter 6 (section 6.2). Based on experience with similar sized plant and also on scientific reasoning and scientific interpolation, the reactor sizing can be done. For example, a typical 50 tpd (15,000 tpa) rotary kiln sponge iron plant has a reactor size of 2.1 m diameter and 30 m length; the corresponding cooler size is 1.5 m Φ and 15 m length.

13.1.4 Process and Material Flow Diagrams

Since the successful process flow sheets roughly correspond to the concepts presented in Fig. 2.2, 13.1 and 13.2, to prepare a fairly detailed process flow diagram is not a difficult task. Thereafter a material flow diagram needs to drawn separately or may simply be superimposed. The material flow diagram emanates from the Material Balance as also from the Energy Balance (may refer to Fig. 2.5 & 2.6)

The importance of Material and energy balancing has been stated in section 7.1. In addition to that if one does not make prior material and energy balance calculation there is a chance that we may end up with unaccounted accumulation and unexpected shortages which may require additional investment to correct for in future.

13.1.5 Preparation of List of Major Equipment

At this point the entire process design has been conceived although it may not be frozen. The designer now prepares the "System Specification of Major Units" in the plant which details the equipment requirement of the unit and their interrelationship. Simultaneously the "Sizing Specification of Major Units" is also prepared which meets the needs of the project. This exercise also helps in estimating the requirement of utilities and power.

The next step is listing out the major equipment and item and indicating their major design parameters. Some typical examples of these major equipment can be found in section 13.4.

13.1.6 Preparation of Layout

From prior experience the process developer lays out the equipment in his mind at various levels and then makes on paper a conceptual block diagram of the layout at different levels of importance. For some of the key regions he makes the elevation block diagram also. If he does not find any inherent conflict he freezes his concept. However, it is easier said than done. Conflicts keep on surfacing again and again and that makes the task of making a layout a long drawn out process. Roughly at this point, i.e., on finalisation of layout, the process design activity ends.

13.2 PROJECT ENGINEERING

Once the process and technology concepts have been frozen and an agreement or an

Table 13.1 Steps in process design and engineering

Sl. No.	*Steps*	*Details*	*Remarks*
1.	Evolution of conceptual design based on input information	Inputs required: • Labouratory investigation • Modelling • Bench and pilot scale validation • Data available in literature on comparable processes • Fundamental aspects	Conceptual design is generally represented in the form of a schematic diagram supported by a short technical write-up
2.	Evolution of basic design, involving: i) Specification of inputs and outputs ii) Flow schemes iii) Plant layout	• Preparation of plant and Equipment layout • Evolving specificaions of raw materials, poducts and other inputs and outputs • Preparing process / material flow diagram	Layout is prepared either • Mentally, or • To-the-scale skethces, considering levels of various equipment and facilities • Area requirement is worked out • Layout is checked for possible interferences, maintenance and safety aspects
3.	Basic engineering, involving i) Preparation of list of equipment ii) Estimation of tonnage/ volume of work iii) Estimation of capital cost iv) Estimate production cost at optimum level v) Prepare activity schedule & Bar chart vi) Estimate cash requirement vii) Prepare estimated profit and loss (P&L) statement for each of the years of construction and operation viii) Estimate internal rate of return (IRR) and the net present value (NPV)	• Volume of work is to be estimated under the heads a) Equipment/plant & Mechinery b) Refractory c) Concreting/Civil work • Capital cost is to be estimated using relevant data, e.g. a) Land requirement b) Soil type c) Land development needs d) Extra facilities required e) Cost of civil work, structures, equipment, installation, commissioning, etc. • Prepare cash flow statement for a) Construction period b) First ten years of operation • Estimate separately for each of the 10 years of operation a) Percentage capacity utilisation b) Maintenance expenses c) Addition, modification and construction (AMC) expenses (if required)	• Most of the inputs to project feasibility report would be estimated or evolved at this stage. • Availability of infrastructure support is a key factor in selection of site. Site accessibility by rail and road, availability of water, other utilities and services, electrical power, etc. would help in reducing the cost of development of infrastrucutre • Activity schedule should be presented in the form of a bar chart • Cash flow in an industry is like the flow of blood in a human body and meticulous estimate is needed to preclude chances of cash shortage • IRR and NPV are considered the most important financial parameters for the evaluation of a project
4.	Preparation of project feasibility report	Detailed in the next Table	Project execution and implementation follow, ending with commissioning and guarantee tests

understanding has been reached with the entrepreneur, a Project Feasibility Report is prepared. The entrepreneur may prepare it himself, may ask the process or technology supplier to do so or may hire an independent Design Consultant. The report may possibly be prepared in two stages. The preliminary (Pre-feasibility) report may focus only on the technical feasibility and the economic viability of the project, while the detailed report (Detailed Project Report or DPR) would additionally have plant and project details with process design figures. These reports often help the entrepreneur or promoter to raise capital from other investors and lending institutions. Table 13.2 gives an outline of items required in the preparation of a feasibility report.

Thereafter Technical Specification (TS) is drawn for tendering purpose. A major portion of the tender specification is the preparation of Detailed Engineering Drawings of equipment, system and other facilities of the proposed plant. It has to be decided beforehand whether the work order for execution of the project would be given in parts or as turnkey contract. This decision would be based on the available expertise – both in open market and in-house – and an assessment of their relative merit. Depending on whether it is part-wise contract or turnkey, the manufacturing drawings may be prepared by the equipment supplier selected, or by the Design Consultant, or by the Technology Supplier. Thereafter the steps to follow are civil and structural construction, equipment construction, installation, cold testing and complete trial run with all interlocking and sequencing. The process finally culminates in hot testing, commissioning, guarantee tests, post commissioning stabilisation, etc.

13.3 INSTALLATION, COMMISSIONING AND PROJECT COST

The gestation period of a DR plant and cost of the overall project depends upon the infrastructural facilities already available, proximity to rail/road/sea connections, cost-cum-availability of equipment and facilities, technology type, expertise, know-how, availability of raw materials, process requirements, etc.

During commissioning and testing period, trials are carried out both under load and no-load conditions. Standardised coal (within a fixed range of sizes, reactivity, ash softening temperature and composition) and iron ore (having specified decrepitability, composition and reducibility) are chosen. Close monitoring of all equipment and systems need to be done during commissioning and post commissioning stabilisation.

13.4 PROCESS ENHANCING EQUIPMENT AND CONCEPTS

With the present emphasis on improved energy efficiency, higher productivity, better product quality and better environment compatibility, a number of optional facilities have been developed by various institutions to augment the existing system for 'high performance DR plants.' Some of the innovations are gas sealing system, preheating system, under-bed hydrocarbon injection system, coarse coal throwing or slinging system, external coal devolatilisation, etc.

13.4.1 Gas Sealing System

It is the system of preventing air-ingress into the rotating kiln, which has made the RK-DR process of sponge iron making technically successful. Earlier attempts were concentrated

Table 13.2 Outline of Preparation of a Project Feasibility Report

Sl. No.	Contents	Details	Remarks
1.	Title and Introductory section	Names of • Project • Report preparing agency • Sponsoring organisation Statement of purpose	Name should give an idea of contents Reference should be given to the relevant interaction with the sponsoring organisation
2.	Executive summary	A synoptic view of all the essential findings of each chapter	Balance to be maintained between brevity and content while maintaining readability
3.	Project background	• Details about project sponsores; their strengths and weaknesses; relevant experience • Project history • Introduction to the process • Studies and investigations already performed	Briefly state • Conclusions arrived at and decisions taken from these former studies; • Pre-investment studies, e.g., Opportunity studies, Land surveys, etc.
4.	Market and plant capacity	• Study of demand and market of the product • Sales forecast and marketing • Production programme • Determination of plant capacity	Describe: • Factors taken into consideration for going into this project • Existing size and capacities of the industry • Its part growth and the estimated future growth • Major problems and prospects • Sales forecast and marketing • Anticipated competition • Quantitative relationship between sales, plant capacity and material inputs
5.	Material inputs	Describe: • Input requirements • Their present and potential supply positions • An estimate of annual costs of local and imported materials	Inputs to be listed separately under the following heads: • Raw materials: • Processed industrial materials; • Components; • Auxiliary materials; • Factory supplies; • Utilities, especially power
6.	Process/Technology selection	Analyse: • Alternative routes and processes available • Criticial study of alternatives • Process selection with key criteria for selection	Credentials of prospective technology suppliers, their background and experience, are also to be analysed
7.	Site selection	Pros and cons of alternative sites and reasons for selecting a particular site to be stated	Land requirement and its cost also to be stated

(Contd...)

Sl. No.	Contents	Details	Remarks
8.	Project engineering	• Statement of scope • Plant Layout, Design and Description • Equipment selection • Raw material sourcing • Utilities and Auxiliaries • Instrumentation and control equipment • Civil engineering works • Construction Schedule	
9.	Plant organization and overhead costs	Estimate: • Plant operation schedule • Organisation structure • Manning and Training • Overhead costs	Overheads based on: • Factory overheads • Administrative overheads; • Financial overheads.
10.	Manpower	• Estimate requirements based on organisation structure • Estimate manpower cost based on salaries	Managerial, executive, supervisory, skilled and unskilled manpower should be separately identified
11.	Implementation scheduling	Should include: • Implementation time schedule • Implementation costs	• Bar Chart is necessary • PERT Chart to be included for larger projects
12.	Financial and economic evaluation; comprising: • Investment costs • Project financing • Production cost • Economic evaluation	Based on: • Investment costs • Working capital requirement • Capital structure • Fixed and variable costs • Financial evaluation	Should include: • Pay-off period • Simple rate of return • Break-even point • Internal rate of return • Sensitivity analysis

on the draught control and maintaining slight positive pressure in the reactor to achieve this end. Subsequently, mechanical seals were tried which followed pumping of temperature resistant grease (High Temperature Grease) into the surface of contact. The present widely accepted system of sealing is schematically represented in Fig. 13.4 and comprises two mechanical slip ring plates in contact, one welded to the rotating shell of the kiln while the other, with suitable mounting device, slides over the stationary hood. The sliding surface with the hood is sealed by means of asbestos pad pressed with mechanical springs while the sliding slip ring plate is pressed by means of pneumatic cylinders against the slip ring plate fixed to the kiln shell. High temperature grease is pumped continuously in between the two plates of slip rings for the dual purpose of lubricating the contact surface and sealing any small crevices that would otherwise cause gas leakage.

13.4.2 Underbed Hydrocarbon Injection System—The Ported Kiln Concept

The system of gas sealing described above was developed by M/s Lurgi, Germany as part of the SL/RN process, (developed jointly by Stelco (Steel Co. of Canada), Lurgi, Republic Steel and National Lead. In the 70s M/s Allis Chalmers announced the development, first of ACAR Process (Allis Chalmers Agglomerated Reduction Process), and subsequently of ACCAR Process (Allis Chalmers Controlled Atmosphere Reduction Process), which claimed many additional advantages. They used a ported kiln concept with a large number of tuyeres in the reduction zone of the kiln. The same tuyeres injected alternately air and hydrocarbon depending on whether it was passing through the freeboard (gaseous zone) during rotation of the kiln or was under the solid charge bed (Fig. 13.5). This was the overbed air and underbed oil injection system and was made possible by the rotary valve - striker arm concept. In larger kiln the valve-striker system has to be supported by a pneumatic triggering device and solenoid valve system.

The overbed air injection is a necessary part of the process (Chapter 2) so that controlled oxidation of the CO in freeboard helps in maintaining a near constant temperature profile over the entire reduction zone of the kiln. In the Lurgi concept this

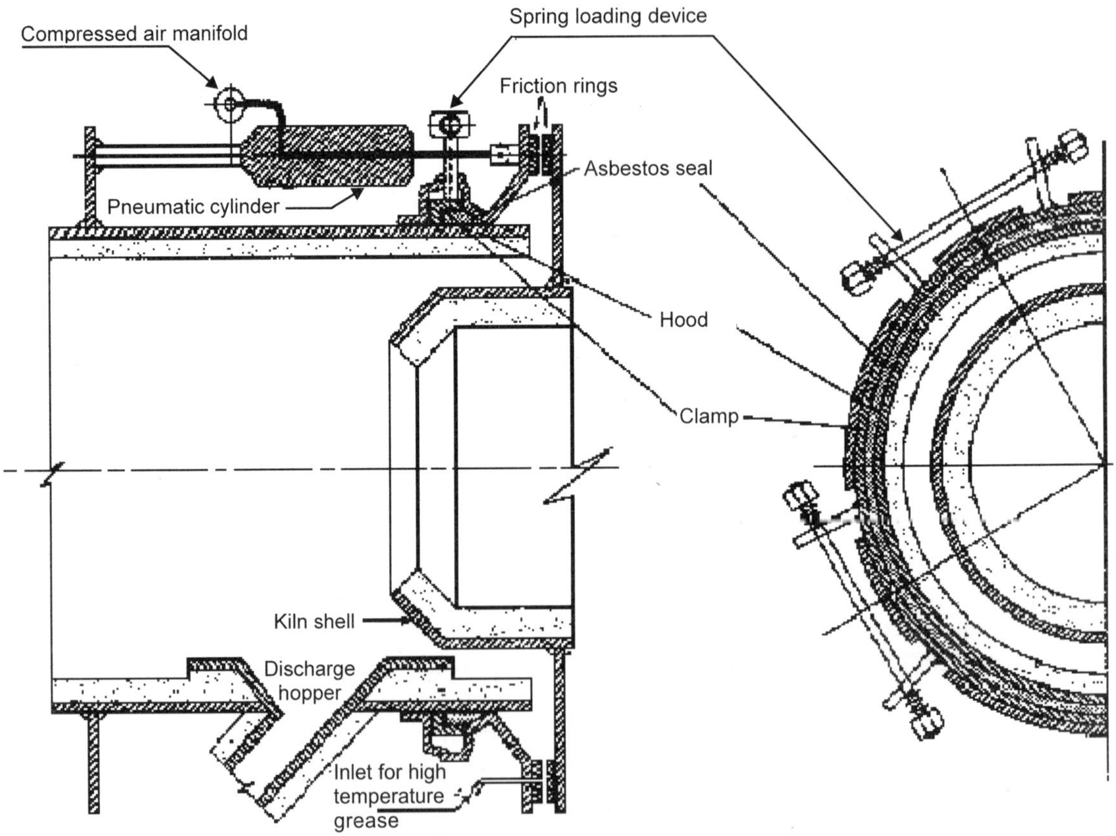

Fig. 13.4 Schematic representation of a slip ring gas sealing device

requirement was met by provision of air-tubes, which are temperature resistant permanent metallic tubes, which release the air from the shell mounted air fans along the axis of the kiln, so that the air released does not oxidise the reduced iron in the charge bed. The Lurgi concept does not require the complicated rotary valve-striker arm, pneumatic trigger and solenoid valve system. On the other hand, the air tubes use up a part of the inner volume of kiln and makes the process ever so slightly more accretion prone. Experiences of operating kilns have shown that the perennial problem of accretion or ring formation generally starts from the base of such air tubes. And in converting existing cement or calcination kilns to sponge iron kilns, diameter is generally a constraint, and by utilising the ported kiln concept and valve–striker arm device rather than air tubes, one can save on the inner volume available to the process.

But Allis Chalmers originally planned the system to inject hydrocarbons while the tuyere passed under the solid charge bed during rotation of the kiln. They claimed, probably rightly, a higher productivity, a lower temperature of reduction (i.e., lesser propensity to accretion formation) and a better product having higher carbon content and a generally higher level of metallisation.

The plant of M/s Orissa Sponge Iron Limited (OSIL) set up at Palaspanga in Keonjhar District in eastern India, was the first and only commercial plant set up on the ACCAR process concept. The commissioning trials were started in 1982 but the plant could be fully commissioned only in 1985. By that time the increased oil prices made the oil injection process unviable.

Even though OSIL has abandoned oil injection system, they have modified the ACCAR design and has supplied it as OSIL technology for overbed air injection in newer units.

In the Sponge Iron Pilot Plant Complex of RDCIS, SAIL a project was undertaken to parallely develop an underbed oil injection system. A cold prototype was successfully developed which was converted to a hot prototype and thereafter was transplanted with minor modifications on to the Pilot Plant. Instead of transplanting the system on the kiln it was installed in the beginning of the rotary cooler, where the temperature of sponge iron was still very high. Up to the point of oil injection, the cooler was refractory lined to retain the temperature in the solid bed.

The outcome of the development, even though was technically satisfactory, the benefit obtained could not justify the expenditure on the quantity of oil injected.

Even though the underbed oil injection system had been a new concept evolved in the 70s, the tuyere design existed earlier and was used for underbed steam injection for selectively reducing Fe_2O_3 to Fe_3O_4 in lean iron ores, so that the product could be beneficiated by magnetic separation (the magnetising roasting method of beneficiating lean iron ores).

13.4.3 Coarse Coal Throwing or Slinging System

At the charging end of the rotary kiln, granular iron ore (generally 5 to 25 mm size) is charged along with coarse coal and flux. Fine coal (0 to 3 mm) is generally not charged mainly because the ultra-fines therein tend to get carried away along with the waste gas. Apart from the energy lost this way, the fine coal suspended in the hot gas presents a safety hazard as well.

Towards the latter half of the kiln, where FeO is being reduced to metallic iron, the char in the bed does not remain as reactive as it was when freshly charred. But it is in this part of the kiln that more reactive char is needed since the final stages of reduction are the most critical ones.

To supply this reactive char, almost all variants of the process throw (or sling) a small (or even larger) percentage of coal as fine coal from discharge end. This throwing or slinging is done with the help of air as medium. Even though the quantity of air used is only a small fraction of what is needed for complete combustion of this coal, the transportation or throwing is achieved in 'Dilute Phase' (as against Dense Phase Transportation of Pulverised Coal and other finely ground material in many applications).

While the principle of throwing appears quite simple, there are a number of precautions to be taken while undertaking throwing of coal. Further, there are a number of limitations with coal throwing as well. These have been detailed in chapter 6 and we summarise below some of the important points.

Fine coal particles cannot gather enough momentum to reach sufficient length of kiln by themselves; it is the current of freeboard gas moving towards the charging end, which "shoulders" the fine particles to reach appreciable distances.

The coarser particles can gather sufficient momentum to reach larger distances. But for this the particles should be given sufficient time and distance to attain full momentum before being released into the kiln environment. OSIL plant utilises an optically aligned long pipe, fully straight, where the 6 to 25 mm size coal is carried by air for full 35 metres or more before being 'slung' into the kiln freeboard. This way they have been successful in throwing from the discharge end 60 to 70% of the total requirement of coal. The more we use coal from the discharge end the more energy efficient the process would be as the volatile matter in the coal would be utilised within the kiln itself. From the feed coal, the volatile matter joins the waste gases and are not utilised within the kiln.

In our zeal to throw more coal from discharge end we may be sending in more and more of air (as carrier) as beyond a limit, the air may disturb the delicate balance that is maintained between oxidising and reducing layers in the rotary kiln (Fig. 2.9). Too much air with the coal may lead to oxidation of the already reduced sponge iron resulting not only in loss of metallisation, but also rise in temperature and chance of accretion formation. From this one can conclude easily that in a larger diameter kiln one should be able to throw a higher quantity and percentage of coal without problem.

13.4.4 Waste Heat Recovery System—Generation of Electrical Power

In Fig. 2.6 and Chapter 7, it has been emphasised that a large part of the energy, put in as coal, goes out with waste gas–both as sensible heat and chemical energy. The plant of M/s ISCOR at Vanderbijl Park in South Africa was the first plant to recover this energy to generate electrical power.

In India M/s Jindal Steel and Power Limited (JSPL) located at Raigarh in the central Indian state of Chhattisgarh has also installed a large thermal power plant (150 MW) which is based on energy from waste gases from ten modules of rotary kiln. M/s HEG Limited in the same state, located nearby at Durg, also generates thermal power using heat from waste gases from just two kilns, each of capacity 30,000 tpa. The quantity of electrical power generated by both these plants are well in excess of their internal requirement, and the surplus power is being sold to the state electricity grid. The viability of this concept has been possible because of remunerative repurchase arrangement with the state electricity board. This has also helped the state in overcoming its electrical power shortage.

13.4.5 Use of Waste Gas Energy for Preheating

Even though power generation from the energy in waste gas is one feasible option for utilising waste energy, it is always preferable to plough as much of this energy back into the process as possible. Some plants do use the fully burnt off gas for drying of raw materials with significant benefits, but this process can use only a fraction of the energy available. And since drying anyway takes place in the beginning of the kiln, the drying device as mentioned above, amounts only to some extension of the kiln volume.

The Tata Steel's development, the Tisco Direct Reduction (TDR) process originally envisaged in-situ pelletisation of iron ore and induration of green pellets by the burnt waste gases. A prototype was installed but there were difficulties with initial trial runs. After it was realised that pellets are not going to be viable for the rotary kiln process, the concept was abandoned.

The waste gases are highly dust laden, mostly with fine ash from coal as also with fine iron ore and fine coal particles, generated during tumbling within the kiln. A shaft reactor containing granular iron ore could have been used to recover the heat from waste gases, but for the suspended dust, which would choke the bed voids, thus preventing further passage of gas. Partial cleaning of gas beforehand would require cooling of the gas and would therefore defeat the purpose of the system.

Heating of iron ore in another rotary kiln is an attractive possibility. Since preheating of coal would lead to loss of its volatiles, coal should not be preheated beyond the drying range. At RDCIS SAIL a heat transfer model

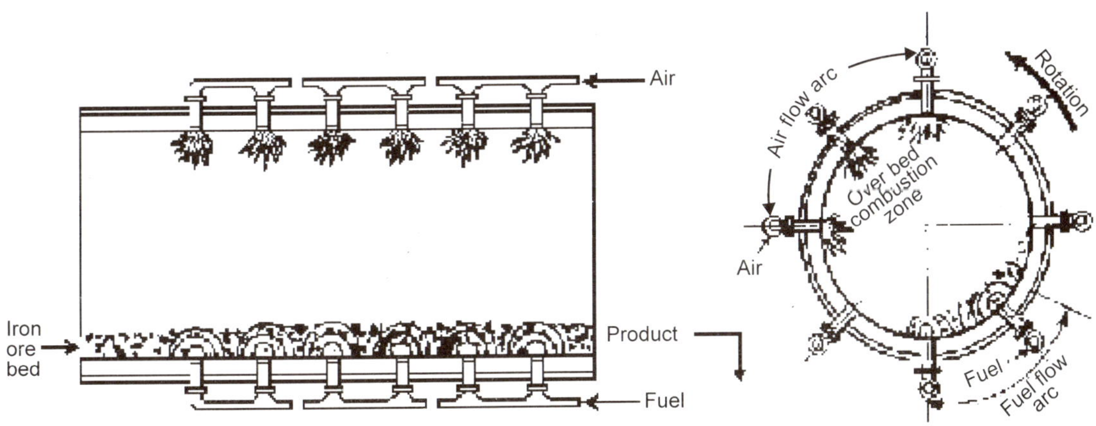

Fig. 13.5 Schematic of ACCAR process ported reactor

was developed based on the principles presented in Chapter 7. The model was validated with the data collected in its pilot rotary kiln when it was run in the preheating mode. Based on the model the following observations were made:

1. For effecting the same level of heat transfer the length requirement of kiln increases only marginally with decreasing diameter. As a thumb rule the length requirement may be assumed constant.

2. With increasing filling degree of the kiln, the efficiency of heat transfer increases. Since there are practical difficulties in operating the kiln beyond 45 to 50% filling degree, and also because substantial freeboard is required for easy passage of gas, a filling degree of about 40% is considered optimum for a rotary kiln to be an effective recuperator.

In contrast to the above, for a reduction kiln, the filling degree must be much lower. Practice has shown that higher than 15% filling degree impairs the process of reduction. This may be because for reduction purpose a large free board is needed to maintain the delicate balance between oxidising and reducing zones.

Thus the design of the sponge iron rotary kiln is a compromise between the requirements of reduction and heat transfer. If the two functions are separated out in two different kilns then there need not be any design compromise, and the entire function can be carried out in an overall lower kiln volume.

For a 500 tpd plant using coal with 6.4% moisture and 24% ash, the dimensions expected for the reduction kiln are 5 m Φ and 40 m length, and for the preheating kiln these are 3 m Φ at discharge end and 30 m length. The preheating kiln, where a large volume reduction of gas is expected due to temperature drop, may take a conical shape for greater effectiveness, in which case the charging end diameter would be 2.1 m. The diameters given above are all unlined diameters. Refractory lining and insulating layers together are expected to occupy a thickness of 250 mm and the lined inside diameters would be lesser by 0.5 m.

The above analysis has led to the concept of an optimised rotary kiln sponge iron making process, which was mentioned in Chapter 5 and is described in the next section.

13.4.6 An Optimised Rotary Kiln Sponge Iron Making Process

Figure 13.6 schematically represents the optimised combination of equipment for running a low energy consuming rotary kiln sponge iron plant. As mentioned earlier, the preheating kiln is so designed as to heat the iron ore to about 700°C or more using the energy in the waste gas. The discharge end of this preheating kiln, where the waste gases enter in counter current fashion, and the combustibles are burnt with preheated air, effectively works as the After Burning Chamber (ABC – also called After Burning and Cooling Chamber) and no separate ABC is required. The rest of the preheating kiln effectively acts as a cooler for the gases and its temperature should come down to about 500°C.

Along with iron ore, the flux and recirculated char can also be heated. This way flux would be calcined and the recirculated char would get ground since iron ore would act as the grinding medium. With fresh surface of char being exposed, gasification would be easier. The low reactivity of char deters its recirculation in the normal process, but this way not only would the char be rendered reactive and useful, there may be enhanced heat transfer in the preheating kiln due to partial reduction of Fe_2O_3 to FeO.

In Chapter 5 we have presented a method of theoretically calculating the coal requirement in sponge iron making process in rotary kiln. Based on this calculation, the expected coal consumption in the normal as well as optimised process can be compared.

In the calculations presented in Chapter 5, four variants are considered. In all variants, waste gases are assumed to be completely combusted i.e. no volatile matter is allowed to go out free or partially reacted. 200 kg/t of char is assumed to be recirculated in the fourth variant. If char addition is not desired it can be replaced by 151 kg of coal.

To recall, the variants considered in Chapter 5 are:

(i) Conventional process, waste gas going out at 1200°C

(ii) Conventional process, waste gas going out at 1000°C

(iii) Conventional process, waste gas going out at 800°C

(iv) Process with preheating of iron ore, char and flux to approx. 700°C (664°C) and air preheating to an average temperature of 400°C.

In variants (i), (ii) and (iii) the system has been considered up to a point where the combustibles in the waste gases have been fully burnt with stoichiometric amount of air, but the heat so developed has been assumed to be still retained by the gas in the After Burning and Cooling (ABC) Chamber. In actual practice, burning and cooling zones (cooling by water spray) cannot be sharply demarcated. Rapid transfer of heat starts before completion of combustion. As a result the temperatures in actual ABC's are much lower than what is indicated here. This does not affect the results of this calculation. However, it is difficult to ascertain with which variant an actual process needs to be compared. This is the reason why three similar variants, differing only in the theoretical maximum temperature of ABC, have been presented.

The processes presently commercialised in India would generally correspond somewhere between variants (i) and (iii).

In variant (iv), inputs have been taken from a heat transfer calculation carried out based on a model developed and tested in the pilot rotary kiln, operating the kiln as though it were a preheating kiln only. These calculations indicate that, for a module producing 500 tpd sponge iron, a preheating kiln of 3 m dia (2.5 m with refractory lining) and 30 m length, and preheating charge to 664°C would mean waste gases going out at 700°C. This is without any drying of charge, reduction of iron ore or calcination of flux taking place in the preheating kiln. If a partial reduction and/or calcination takes place, heat transfer would be further enhanced and energy saving would be still better. Even traces of CO and/or H_2 are adequate to reduce Fe_2O_3 to FeO, and this is highly feasible in the preheating kiln. Outlet gas temperature should come down to about 500°C.

A 500 tpd kiln of conventional type is likely to have dimensions 5 m Φ * 70 m length. In the preheating zone, length required for the same extent of heat transfer increases very marginally with decreasing diameter. Thus, when preheating kiln is used, a possible combination of dimension would be

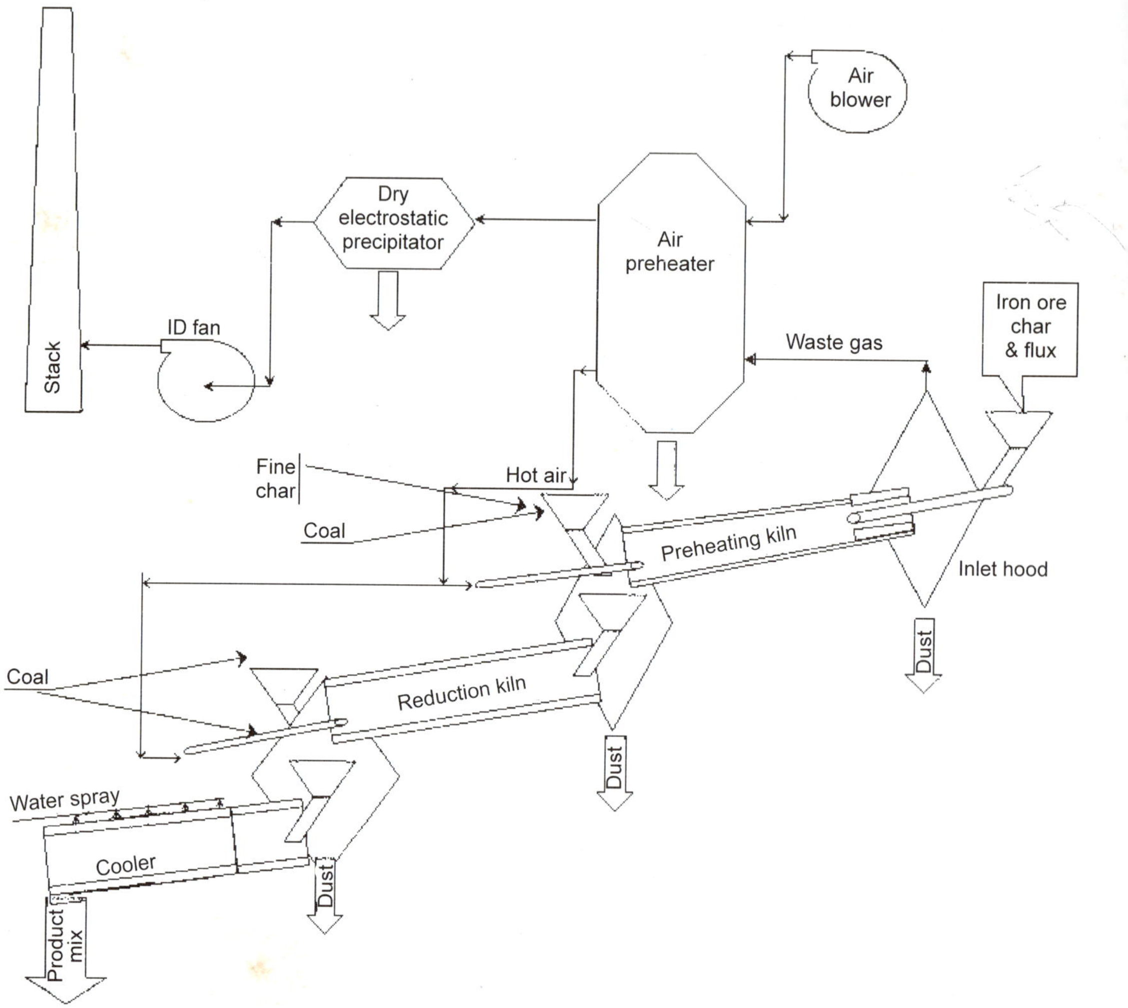

Fig. 13.6 An optimised rotary kiln sponge iron making process

(a) Preheating kiln 3 m Φ * 30 m length and

(b) Reduction kiln 5 m Φ * 40 m length

Tables 5.3 to 5.9 present the method of calculation of coal requirement of sponge iron rotary kiln for the four different variants. In Table 5.9 we can find the heat deficits for these alternatives and thereby the total coal requirement.

One simplifying assumption which has been made is that the average molar specific heat of waste gases has been assumed to be 9.0 kcal/kg mole - °K. One can repeat the calculation with the actual Cp values of the constituents, which are functions of temperature, and integrate over the relevant temperature range to obtain a more accurate result.

Figure 13.6 represents a possible plant based on the optimised concept. The diagram has been made keeping 500 tpd plant in mind although it could well be applicable for a 50 tpd or even a 25 tpd plant. Essential equipment have been presented in the diagram and it can form the basis for preparing detailed flow diagram and the equipment list.

13.4.7 Pros and Cons of the Optimised Concept

As mentioned earlier, for utilising the energy in waste gases, many of the existing plants have resorted to electric power generation. While this is a viable option for large plants (of capacity) 60,000 tpa and above, but for smaller plants, such power generating units are not viable. The result is that many units of capacity 15,000 tpa (50 tpd) have come up and are operating off and on as they are finding it profitable to run only during the periods of good market demand. Over the years, they have not been able upgrade the process to make it viable during the depressed market conditions or are unable to generate enough funds for installing additional units to a capacity level where installation of an electric power-generating unit would be viable.

With the optimised process concept, maximum possible energy would be utilised within the process with marginal extra installation cost. Operating cost would reduce drastically since coal consumption would be halved. The After Burning and Cooling function would be conducted inside the preheating kiln, which would also carry out the function of grinding of char for complete utilisation of its carbon content. Only the Air-preheater would be an extra installation. But this would also be in lieu of the Air Cooler or Gas Conditioner which are presently needed to make the waste gas suitable for being treated in the dust removal unit (electrostatic precipitator or bag filters).

Table 13.3 gives a comparison of the optimised process with the conventional process. The optimised process can be seen to be much more environment friendly, besides other benefits mentioned earlier.

13.5 KILN REFRACTORIES

In direct reduction kilns, selection of proper refractory material does not pose a serious problem mainly because of relatively low operating temperatures. Dense-fired alumino-silicate refractories in the range of 45-85% alumina should meet the requirements, provided they are resistant to abrasion and to highly reducing conditions. Normally Andalusite based firebricks with low thermal conductivity (1.0 w/mk) and high cold crushing strength (100 N/mm^2) are preferred. However, the experiences of the operating plants are more in favour of using high alumina castables. Even though it calls for higher investment initially, performance has so far been found to be much more reliable.

The reason for the higher reliability of castable refractory may be due to the better resistance it offers to thermal fatigue. Schnabel (1977) estimated that refractory surface gains a temperature of about 80°C during the time it is exposed to free-board; and loses it when it is in contact with the solid charge. Since this happens with each revolution of the kiln, one can imagine the extent of heating and cooling cycles to which the refractory surface is exposed.

For using castables, the inner shell of the kiln has to be welded with large number of evenly spaced stainless steel anchors of 'U,' 'V' or 'Y' shape. Silica wool or similar insulating blanket of 25 to 50 mm thickness is thereafter laid out against the shell followed by 50 to 75 mm layer of insulating blocks. Finally, 125 to 150 mm layer of castables are laid out and then dried and cured as per instructions specified for the castables.

While the three layer refractory lining is the ideal refractory design using castables, most of the plants in India have not adopted this design. Instead they have a single layer of castables with no effort at heat conservation. Consequently the surfaces of kiln shell reach temperatures of 400°C or more while the three layer design of same total thickness could give shell temperatures as low as 100°C or even less.

Reason behind the non-adoption of the three layer design has been the refractory failures encountered in the early indigenised units in India. The cause of these failures could be a combination of poor quality of insulating layers, poor workmanship in laying the individual layers and a faulty anchoring design which allowed a very slight relative movement between the main face castable (i.e. the outer layer exposed to reacting materials) and the kiln shell during the rotation of the kiln. Even the slightest relative movement between these two layers causes the slight grinding action on the insulating layers, lying in between, with each revolution. Under such a condition, before a year is completed, the insulating layer converts into powder and stops giving back-up support to the main face layer. The main face layer thus becomes vulnerable to its own weight and the weight of the material in the kiln.

But the increased radiation loss is an avoidable loss to the process and it is surprising that serious efforts are not being made to make the insulating layer concept a success. One reason could be that the additional radiation loss represents only about 0.2 GCal/tonne of energy loss, which represents only about 3% increase in coal consumption. Doing maintenance work around a kiln whose outside temperature is about 400°C is not easy, but apparently the operators have got used to these conditions.

Table 13.3 Comparison of the conventional and optimised process

Process → Items ↓	*Conventional*	*Optimsed**
Process Parameters		
Energy consumption	5.5 to 6.0 GCal/t-DRI	2.7 to 3.0 GCal/t-DRI
Coal consumption [Coal of 10% Moisture, 25% Ash; Wet basis]	1100-1200 Kg/t-DRI	540-580 Kg./t.DRI
Loss of energy in waste gases	2.0 to 2.5 G Cal/t-DRI	0.35 to 0.40 GCal/t-DRI
Sensible energy loss in solid product	0.28 to 0.32 GCal/t-DRI	0.2 to 0.22 GCal/t-DRI
CO_2 emission	2.1 to 2.3 t/t-DRI	1.05 to 1.15 t/t-DRI
Design Aspects		
Preheating kiln	Not present	Present
Shape of preheating kiln	Not applicable	Conical
Air preheater	Not present	Present
Electrostatic precipitator	Generally present	Present

* This is an improved version of Indian patent No. 170432 (SAIL-DR Process—January 21, 1987).

There is one benefit the operators gain in this situation. If the surface temperature at any time falls sharply below 400°C (measured either by an infra-red camera or hot chalk or a sliding contact thermocouple), they conclude that accretion has formed at this point. If the temperature goes sharply up, they conclude that refractory layer has fallen off.

But a 3% continuous and extra coal consumption is not something which should be ignored. After overcoming defects in quality, workmanship and design, the insulating layer may well be re-introduced in future plants.

There is a parallel development taking place in refractory area, and that is the development of higher strength and abrasion resistant insulating castable. Once such a material is available at reasonable price, the entire lining could just be of this single layer. This would not only mean reduced weight (insulating refractories are highly porous and thus have very low block density) but lower thickness requirement resulting in further reduction in weight.

High strength and abrasion resistant insulating blocks are already available and may possibly be considered for use. But seeing the history, entrepreneurs may have more faith in castables and would, therefore, wait for the development of high strength and abrasion resistant insulating castables.

14 CHAPTER

Other Uses of Rotary Kiln for Reduction Purposes

Rotary kilns are used for a variety of purposes like cement making, and roasting and calcination of lime, dolomite and other refractory materials. While these are examples of oxidising application there are a few examples of reduction in rotary kiln other than sponge iron making.

14.1 MAGNETISING ROASTING OF LEAN IRON ORES

In Chapter 13 the magnetising roasting of lean iron ores have been mentioned. India is fortunate to be endowed with very high-grade hematite deposits. But many other countries of Asia and Europe are not so fortunate. The lean iron ore deposits need to be beneficiated to 65% plus iron. After sufficient size reduction to effect a reasonable liberation of iron oxide and gangue, the ore is treated in a rotary kiln where a mildly reducing condition is maintained. To prevent reduction beyond the desirable level, superheated steam is injected under the bed, using the ported kiln concept as depicted in Fig. 13.6, so that formation of Fe_3O_4 is ensured.

After cooling, the product is subjected to magnetic separation. The magnetite concentrate is collected in the ferromagnetic fraction.

In India, high grade iron ore is in great demand and the deposits are being depleted at a very fast rate. To find iron ore lumps containing 64% or more of iron, suitable for sponge iron making, has become difficult. On the other hand, there is accumulation of lower grade overburden at mining sites and also lower grade fines at production as well as sizing facilities. A good way of removing the accumulation would be to beneficiate them.

Even though the beneficiation plants presently operating and those planned do not have or contemplate having facility of magnetising roasting, but there is a distinct possibility that as the available grades decrease further, this particular concept may re-emerge and be contemplated for installation.

14.2 PRE-REDUCTION OF CHROMITE ORE

In the natural chromite ore, chromium occurs in various states of oxidation, starting with Cr^{3+} to Cr^{6+}. While using this ore in electric arc furnaces for the production of ferrochrome and other chromium based alloys, costly electric power is needed for meeting the endothermic heat required during reduction of the various oxides of chromium. A lot of this electric power can be saved if the ore is pre-reduced using coal and other reductants in a rotary kiln. Ore would get preheated in this process and if charged hot, the sensible heat would help in further saving of electric power.

In conditions similar to sponge iron rotary kiln, chromite ore cannot be reduced to metallic stage, but all chromium is expected to be brought to trivalent Cr^{3+} or lower state. Even this is of great help in the electric smelting process of making chromium-containing alloys.

M/s Outokumpu, Finland has developed a technology of using the smelter off gas, which is rich in reducing gases – largely carbon monoxide - for preheating and pre-reduction of chromite ore before charging in the electric smelting furnace. The ferro-chrome plant of Tata Steel at Bamnipal in the eastern Indian state of Orissa is a plant based on this technology.

14.3 PRE-REDUCTION OF MANGANESE ORE

Just like chromite, manganese occurs in manganese ore in various states of oxidation, from Mn^{2+} to Mn^{4+}. And like chromite, the sponge iron making conditions can, at best, convert all manganese to bivalent stage – no metallisation of manganese is possible. Before smelting in electric ferromanganese furnace, if the ore is pre-reduced and preheated in a rotary kiln, considerable saving in electric power is expected.

While many ferromanganese plants have been considering adoption of this concept for economising on electric power, we are not aware of actual adoption of this system.

14.4 PRE-REDUCTION OF TITANIFERROUS ORE

The metal titanium has very unique properties, unmatched by any other metal or alloy. The lack of existence of an economic method of production has prevented the widespread use of this metal. While its reactive nature at high temperature has forced the adoption of very special extraction technique, the fact that the ores available on land contain very small percentage of titanium has made matters worse. It is only the beach sands of Kerala and Orissa in the southern and eastern shores of India, which are presently being exploited or considered for the extraction of titanium.

Tonnage wise a large quantity of titanium dioxide occurs in or near iron ore deposits as Titaniferrous Iron Ore. These deposits are rarely exploited for using their titanium content (except for a very special application of charging titanium in iron blast furnaces for protecting the hearth refractory). A concept is however being tried, where this ore would be smelted to produce pig iron (by-product) and titanium as TiO would be contained in the slag. The slag would be further processed for titanium extraction. If this scheme is proved successful, pre reduction of the ore – possibly in rotary kiln – has to be an important step for economising on electric power in the process.

14.5 PRE-REDUCTION OF NICKELI-FERROUS OVERBURDEN

Nickel is extracted throughout the world from sulphide-based ores. India does not have such deposits and therefore India has to import its entire requirement of nickel. It is in this context that the nickeliferrous overburden of chromite ore deposit in Sukhinda in eastern Indian state of Orissa assumes importance. This overburden contains significant percentages of nickel and cobalt and a concept is being pursued where Ni and Co would be reduced in a controlled manner to NiO and CoO and then extracted through suitable leachant. In a sponge iron rotary kiln, nickel and cobalt would be fully metallised. Therefore, to effect a controlled reduction to bivalent stage would require considerable control and R&D efforts. But reduction in rotary kiln is a distinct possibility and appears to be worth pursuing. A technique very similar to that of magnetising roasting of iron ore should be a feasible proposition.

14.6 USE OF ROTARY KILN FOR GASIFICATION OF COAL

As ash in coal adversely affects the operation of all reactors using coal or coke, many attempts have been made to minimize or eliminate this drawback. Coal beneficiation by gravity and similar methods has been extensively adopted. But, in Indian coal deposits, the mineral matter (source of ash) is very finely dispersed in the coal seams. Consequently, significant ash elimination is accomplished only with substantial loss of yield. Hence, successful adoption has been only at places where the rejects are gainfully utilized.

One apparently very attractive alternative is to convert the value in coal (C & H) into a liquid or gas which would leave behind the mineral matter. Solvent extraction of coal has been found technically feasible although it is not economically viable. Here the value in coal is dissolved into an organic solvent and filtered. Full recovery of solvent has not been feasible at the moment and that is coming in the way of its commercialization.

However, coal gasification processes exist which can convert coal to a mixture of CO and H_2 through controlled combustion with oxygen. In most of the faster processes the ash is melted and needs to be treated with a flux to make the resultant slag to flow out easily from the gasification reactor. For high ash coals this step turns the process unviable. Jindals are in the process of installing a coal gasification based sponge iron plant at Angul in the eastern Indian state of Orissa where sponge iron would be made in a shaft furnace using gas from coal made in a Sasol-Lurgi gasifier. The latter would be producing the gas at lower temperatures so that ash remains solid and is discharged as such, i.e., without fluxing. This plant may prove to be the mirror of the future large sponge iron plants to come. However, projections indicate that for smaller merchant DR plants, rotary kiln would remain the more viable option.

One more option is available for coal gasification and that is to carry out the process in a rotary kiln. In the present sponge iron rotary kilns both the reduction process and the gasification of coal are being carried out simultaneously. Then, without the iron ore, why can the rotary kiln not produce a reducing gas from coal? We feel that this is definitely feasible and future researchers should investigate this possibility which has immense potential for India. Unlike the sponge iron rotary kiln, there would be a need to enhance gas and bed interaction through the use of higher filling (hence smaller kiln), use of lifters and under bed injection of air or oxygen using the ported kiln concept. Sponge iron would then be made in a shaft, but rotary kiln will remain an important contributor.

15

CHAPTER

Environmental Aspects of Sponge Iron Making in Rotary Kiln and Future Prospects

The subjects of effect on environment and the future of sponge iron making have been linked together with a reason. Even though the immediate future of sponge iron making is dependent on the usefulness of sponge iron vis-à-vis competing items and the value it offers to society, in the long term sponge iron can survive and flourish only if it offers these values along with minimum damage to environment. Any iron and steelmaking operation causes incremental damage to environment, through the release of noise pollution, release of waste heat, solid wastes (slags, ashes, char, etc.), polluting liquid constituents (Phenol, tar, aniline etc.), particulate matter (dust) and harmful gaseous constituents like CO, SOx, NOx and also CO_2. Extensive attempts have been made to control all the above and keep them within limits which the environment can tolerate and absorb, except for the liberation of CO_2.

Normally CO_2 is not considered a pollutant. But with the rapid industrialization of human society, the rate of net release of CO_2 to atmosphere has increased very rapidly and we are already seeing glimpses of the catastrophic change it can cause.

15.1 CARBON DIOXIDE EMISSIONS FROM THE SPONGE IRON INDUSTRY

The average CO_2 level in atmosphere was 0.0295% at the end of nineteenth century, which increased to 0.035% in 100 years. The greenhouse effect it causes, has led to global warming. The ice cap at North Pole has punctured at many places. The larger ice cap in the Antarctic is also receding. Glaciers feeding perennial rivers are reducing in size and we are already experiencing much more extreme weather conditions.

Unfortunately, we cannot conceive of an economic iron and steel making route which does not release CO_2. The cleaner fossil fuel – petroleum and natural gas – are in limited supply. Sponge iron making using such a fuel and reductant liberates much lesser amount of CO_2 but this route cannot be expected to meet fully the iron and steel needed, nor can it be expected to replace the blast furnace route of iron making. But a variation of this route is expected to be the most environment-friendly and hence the acceptable route of iron and steel extraction of the next century, about which we would discuss at the end of this chapter.

Table 15.1 gives us an idea of the quantity of CO_2 being released in the atmosphere and also compares the Indian Scenario with the global practice. Table 15.2 tries to split the CO_2 emissions between iron making and steelmaking stages and demonstrates that most of the emissions take place in iron making stage only. Scrap based steel making has to remain the favoured option for environment protection, but that has to depend upon the availability of recycled scrap, and therefore, can meet only 30 to 40% of the steel requirement. For making steel through extraction of iron from oxides, DR-EAF is the better option at present.

Table 15.1 Carbon dioxide emissions in steel making g

Sl. No.	Process routes	Global practice	Indian practice
1.	BF–BOF	2.4	3.5
2.	Coal based DR–EAF	1.8	2.0
3.	Gas Based DR–EAF	0.8	1.0
4.	Corex–BOF	2.5	2.5
5.	Scrap Based EAF	0.05	0.05

Please note:

1. The inferior figures for Indian practice at Sl. No. 1 & 2 are mainly due to the higher ash levels in coal and coke.
2. Generation of electricity for the electrical furnaces would normally involve CO_2 generation, which has not been accounted for in this Table.

In international forums, there has been a proposal to put a "Carbon Tax" on the industries. As per this a tax should be levied based on the quantity of CO_2 the industries release into atmosphere. Imposition of this tax has been deferred so far due to the skewed nature of the details attached to the proposals so far, which are weighted heavily against developing countries, but we can safely assume that this tax in some form will be levied sooner than later.

Sponge iron industry has an advantage in this regard. But coal based DR stands far behind gas based DR. This fact also indicates that an opportunity exists for coal based DR industry. If all the energy conservation measures are adopted as mentioned earlier in this book (adoption of conical preheating kiln, air-preheater, char-recirculation etc.) the Figure of 1.8 would come down in the range of 1.1 to 1.2 tonnes of CO_2 per tonne DRI.

To put matters in proper perspective, the maximum amount of carbon dioxide which is released into atmosphere by Indian industries, is the thermal power producing industries, accounting for about 60% of the total CO_2 released. Steel industry (iron making + steel making) comes next at about 10%. Out of this, sponge iron making constitutes less than 2%.

15.2 OTHER POLLUTANTS AND ARISINGS IN SPONGE IRON AND STEEL INDUSTRIES

Table 15.3 lists the estimates of amounts of solid arisings in Indian coal based rotary kiln sponge iron plants as per the present practice. Table 15.4 compares these arisings with similar arisings in blast furnace based integrated steel plants. It can be seen that sponge iron plants are already ahead in utilization of its solid arisings even though the specific quantity of arisings is higher.

Table 15.5 lists the pollutants in exhaust gases of rotary kiln sponge iron plants and compares them with those found in blast furnace stack. As far as the liquid effluents are concerned, virtually no such effluents go out of the plant, except where wet scrubber is used. Effluent quantity in case of wet scrubbers range from 800 to 1000 litres per

Table 15. 2 Carbon dioxide emissions at different stages under Indian conditions

Tonnes per tonne iron				*Tonnes per tonne steel*	
Coal based DR	***Gas based DR***	***Blast furnace***	***Smelting reduction***	***Basic oxygen furnace***	***Scrap based electric arc furnace***
1.8 to 2.0	0.80 to 1.0	3.3 to 3.5	2.3 to 3.0	0.15 to 0.30	0.03 to 0.05*
*Once again, CO_2 emission during generation of electricity has not been considered					

tonne sponge iron. Typical values of suspended solids in such effluents are 32 mg/litre (national norm up to 100 mg/litre); biological oxygen demand (BOD) 20 mg/litre (national norm up to 30 mg/l) and chemical oxygen demand (COD) 50 to 96 mg/l (national norm 250 mg/litre).

Thus we see that, from the point of view of emission of pollutants, as also CO_2 emission, sponge iron making in rotary kiln is better placed than blast furnace iron making. As stated earlier, if all the energy conservation measures mentioned in earlier chapters, are adopted (conical preheating kiln, air preheater, char recirculation, etc.), the CO_2 emission is poised to go down to as as 1.1 to 1.2 tonnes per tonne sponge iron.

Environmental concerns may lead to adoption of zero CO_2 emission technology of steelmaking in the next 100 years about which we would discuss towards the end of this chapter.

15.3 SPONGE IRON IN INDIA

Even though sponge iron making in rotary kiln made significant starts in South Africa, South American countries Peru and Brazil and in New Zealand, India is now the main centre with much more than half of world production concentrated here.

Presently, sponge iron accounts for about 40% of total iron production in India and more than half of it is from rotary kiln. World production figures are not so significant which underscores the position India occupies in rotary kiln sponge iron making.

Table 15.3 Estimated average specific arisings from Indian coal based sponge iron plants

Sl. No.	*Items*	*Quanity*	*Remarks*
1.	Iron ore fines	153	Fully sold or utilized
2.	Kiln backflow	240	Partly sold or utilized; rest dumped
3.	Dolo-char and ash	355	Partly utilized or sold; rest is dumped
4.	Agglomerates and accretion	52	Mostly dumped
5.	Fly ash	300	Average for plants having power plants.

Table 15.4 Comparison of solid arisings from blast furnaces vis-a-vis rotary kiln sponge iron plants

	Total quantity of arisings		*Quantity of arisings recycled/reused and sold*		*Percent utilized*
	Million tonnes per annum	*Kg/tonne steel or sponge iron*	*Million tonnes per annum*	*Kg/tonne steel or sponge iron*	
Blast furnaces (integrated plants)	5.072	479	2.916	275	57.5
Coal based rotary kiln sponge iron plants		800		638	78.5
Does not include fly ash or sinter plant wastes					

Table 15.5 Comparison of pollutants in exaust gases in blast furnace iron making and coal based rotary kiln sponge iron plants

Maximum Levels :
Suspended Particulate Matter (SPM - National Norm) = 150 mg/Nm³
Oxides of Sulphur (SOx - International Norm) = 50 mg/Nm³
Oxides of Nitrogen (NOx-International Norm) = 150 mg/Nm³

mg/Nm³	SPM	SOx	NOx
Coal based rotary kiln sponge iron plants - Exhaust Gas	80 to 120	Up to 110	UP to 25
Blast Furnace Stack Gas	6 to 26	9 to 124	3 to 48

Almost the entire sponge iron production is consumed for steel making purposes. The following projection indicates that steel consumption in India is likely to peak in 2050 AD at about 170 million tonnes.

15.4 PROJECTIONS OF STEEL CONSUMPTION

Steel has so far proved to be the single key factor responsible for industrial production, and thereby, for economic growth. And it is growing from strength to strength with newer developments – both within steel making practice as well as engineering developments, which ask for more usage of steel. So much so, that economic development has become almost synonymous with steel consumption for a developing country like India.

Trend studies have revealed that steel consumption is strongly dependent on the state of development of the economy where it is consumed. In this connection, the economies of Germany, UK, USA and Japan can be considered as developed economies where the consumption had risen to a peak and then declined and now tends to fluctuate around a stagnating curve.

If we plot the normalised consumption of steel (i.e., ratio of consumption to peak consumption value) against normalised time of peak consumption (ratio of years elapsed from a judiciously chosen starting year to the years taken to reach peak consumption) we get a plot of the type shown in Fig. 15.1.

World consumption has not been showing such a trend, or rather the peak has not been reached yet. Since world consumption would actually be superimposition of many such curves in a staggered manner, the peak is expected to be erratic like a mountain range. In Fig. 15.2 is plotted the production figure, which should be consistent with consumption, although with correction factors for yield, stock build-up and releases.

China is presently the dominant player in steel consumption pattern of the world.

Apart from being the largest steel consumer, it is also the largest importer, and the latter has contributed maximum in stimulating steel production all over the world – as also in India.

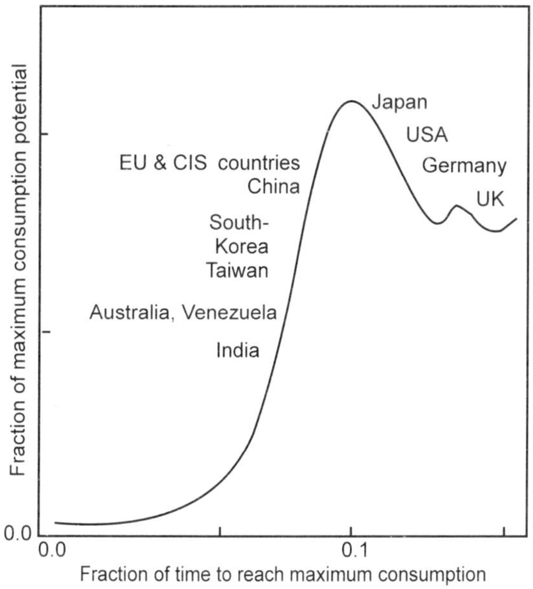

Fig. 15.1 Steel consumption in present day economies

For a long time India's steel production and consumption figures were roughly matching, but in FY 2002-03 India produced about 32.5 million tonnes of finished steel, 3.6 million tonnes more than its consumption, largely because of increased exports. In FY 2003-04 these figures increased to 36.2 and 5.8. In spite of the recession in the domestic steel production thereafter, yearly consumption rate increased consistently. Many times, the actual consumption has been falling short of projected figures. But in 1991-93 period the consumption increased by 21%, averaging 10.5% per year, as against a six-decade average growth of 5.5% per year till that time. The six-decade average growth in the index of Industrial Production was also 5.5%, which suggested direct dependence of industrial growth on steel consumption.

15.5 PROJECTIONS FOR INDIA

In-house projection indicated that India's steel consumption would peak to 170 million tonnes by 2050 AD (Fig. 15.3). This projection was made in 1997 based on trend analysis and the outcome appeared very improbable. But since then the changes occurring in the steel industry of China has raised hopes that this type of change is very much likely.

For this to happen, the steel consumption in India will be increasing by about four times in the next four decades.

15.6 POSSIBLE SOURCES OF ADDITIONAL IRON

For meeting the corresponding increased demand of iron in India, blast furnaces will

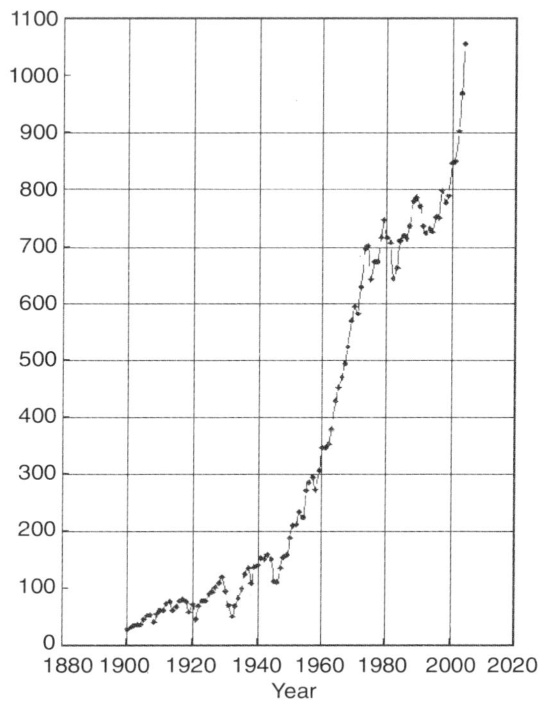

Fig. 15.2 World crude steel production

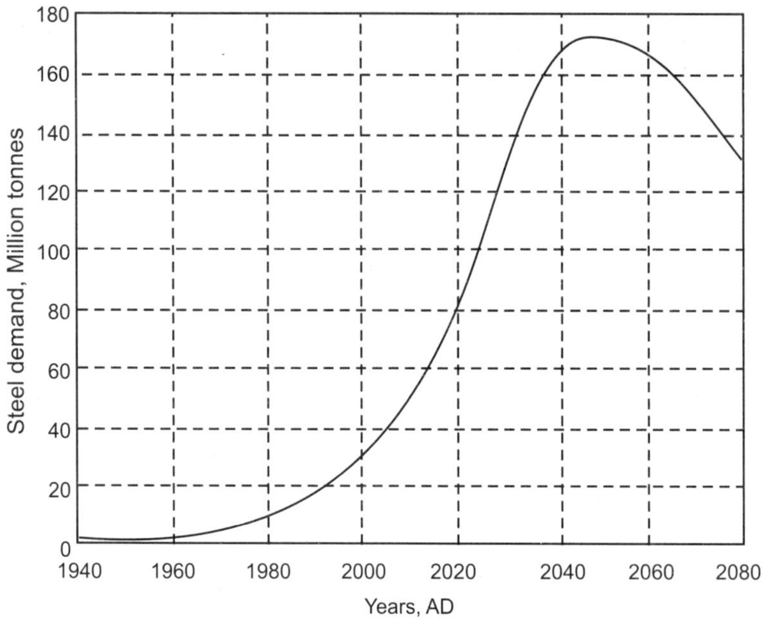

Fig. 15.3 Long term steel demand projection for India

continue to play significant role. But its dominance is likely to be eroded largely because of shortage of metallurgical coal. The availability of the same may remain restricted within the country. Blast furnaces may not only be surviving but the numbers may grow based on imported metallurgical coal. But even these imports may gradually prove cost wise prohibitive. In spite of these problems, mini (and micro – 23m^3) blast furnaces are now proliferating.

About 55% of iron in India is extracted in blast furnaces, while 40% comes from the DRI route (Table 15.1). In India, scarcity of coking coal has been forcing adoption of processes involving non-coking coal. The Indian non-coking coals are unsuitable for many applications, but appear well suited for treatment in rotary kiln sponge iron making processes – well suited for treatment in rotary kiln sponge iron making processes. The blast furnace iron making would stay as long as it can run efficiently – would use less and less of hard coke, may use even formed coke – would inject more of alternative fuels like beneficiated and pulverised non-coking coal, tar, etc.

In Table 15.1, we attempt to predict the situation in India after about 4 decades. The overall global situation would be roughly compatible with the situation in India as regards the contribution of different routes.

15.7 ROLE OF SPONGE IRON AND NEWER PROCESSES

While changes in rest of the world are going to be in similar pattern, the DRI or sponge iron part would be of coal based variety in Indian sub-continent while in rest of the world (with possible exception of African continent) the natural gas based sponge iron would predominate. Smelting reduction processes like Corex, Romelt and/or other newer processes would supplement the hot metal production from blast furnaces. While blast furnaces would be using a minimum amount of hard coke, even this may be supplemented by formed coke or something similar. On the other hand, hard coke may be made differently. There is a promising trend of adopting heat recovery type non-recovery and stamp charged coke ovens. Of the various attempts on making good quality coke from inferior coals, the route of increasing the bulk density of blend (through

Table 15.6 Likely Indian steel making scenario in mid twenty first century

(Inferred from 'Steel Demand Forecasts: India 1999-2000'–Steel Times International, 23 (4), pp. 14-17 (July '99)

(Million tonees)

	Present	*Expected after four decades*
Total steel production	50	170
Steel production by Hot Metal-BOF Route (Contribution of SR-BOF likely to increase from about 7% to about 20%)	30	85
Steel production in Electric Furnaces (Hot metal/Pig Iron + DRI + Scrap) (Contribution of DRI is likely to be maintained at about 70%)	20	85
Matching DRI Production	18	75
Matching BF hot metal	27	65

stamp charging or otherwise) has proved most economically viable. We may see alternatives based on this principle being pursued further.

15.8 ROTARY KILN SPONGE IRON MAKING: STATUS AND PROSPECTS

In India the growth of coal based rotary kiln sponge iron making has been phenomenal. From a modest start in 1980 with demonstration plant of 100 tpd of Sponge Iron India Limited in the southern Indian state of Andhra Pradesh, the Indian sponge iron industry struggled to take off for nearly 10 years. Between 1990 and 1996, three large gas based sponge iron units were commissioned, which boosted India's sponge iron production. The production remained stagnated, although did not decline under adverse circumstances, till the turn of the century. In the 21st century, the coal-based units have got a boost and have catapulted India to the top of sponge iron producers overtaking Venezuela, Mexico and Iran. It is expected that very soon more than 200 coal based rotary kiln sponge iron plants and nearly 400 sponge iron rotary kilns will be operating in India. Presently, sponge iron production in India by the coal-based route is nearly 4 times the rest of the world put together (Fig. 15.4).

However, for sponge iron makers in rotary kiln, the quality of coal, especially its ash content and reactivity, is a real challenge. They are sometimes forced to use 'F' (Moisture + Ash = 40 to 47%) and 'G' grade coals with consequent loss in productivity and quality of sponge iron. Many of the producers are hoping that a suitable coal gasification process would be developed, which would help them in eliminating ash from the sponge iron making reactor. While such a process has been technically proven, economic viability has eluded it so far. Gasification of coal in a rotary kiln itself may prove to be the answer to this problem. No other metallurgical reactor can handle ash of coal better than a rotary kiln. While coal gas based sponge iron would be made in shaft furnaces, rotary kiln may have an important role here as well.

15.9 USE OF SPONGE IRON IN BLAST FURNACES

The implications of charging sponge iron in blast furnace have been covered in chapter 12. We reproduce a summary below, along with projections.

The steel consumption in India is increasing at such a rate that all of the produce from both blast furnace and DR units are fully consumed, and there is demand for more. Therefore, blast furnaces do not as yet see DR units as a threat. In fact, Indian blast

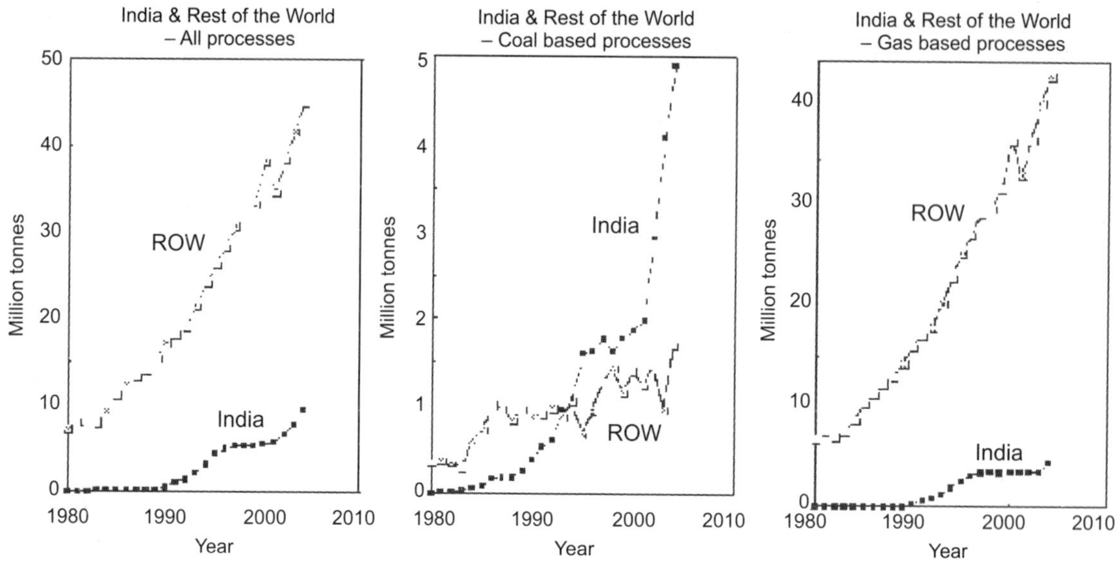

Fig. 15.4 Sponge iron production over the years
(Charts depict the phenomenal growth of coal based sponge iron making in India)

furnaces have started supplementing its feed with sponge iron. Blast furnaces of Tata Steel and one blast furnace each of the integrated steel plants at Bokaro, Rourkela, Bhilai and Durgapur have experimented with regular use of sponge iron. Some of them have touched the usage rate of as much as 180-kg/tonne hot metal. This is being done in spite of the fact that, till hot metal stage, the benefits obtained are always significantly below the cost incurred.

Sponge iron charging in blast furnaces is being resorted to when there is large demand for hot metal within the steel works and the availability of coke and coking coal is substantially reduced. The overall plant economics justifies the use of sponge iron in the blast furnace burden. Projections indicate that the use of sponge iron in blast furnace will very soon reach a plateau and could then start coming down – may even stop. It is expected that after a couple of years the charging of sponge iron in blast furnaces would be started once again when cost of sponge iron goes down to below about 40% of coke price. That time, the use of sponge iron in blast furnace may reach an average level of 10 to 15% of the charge. The demand projection of sponge iron, which is presented in Table 15.1, does not take into account the possibility of this consumption of sponge iron in blast furnace. Use of sponge iron in blast furnace at low rates is expected to boost both productivity and coke saving by 1% for each percent metallisation of the burden. Under Indian context with use of coal based sponge iron, the values come to 8% of coke saving and as much percent of productivity increase for 10% metallization of burden, corresponding to about 9% sponge iron of the total normal iron oxide in burden.

15.10 SMELTING REDUCTION IN INDIA

Contribution by smelting reduction (SR) processes to iron making is much less compared to both blast furnace and DR. But we would like to make a special mention of

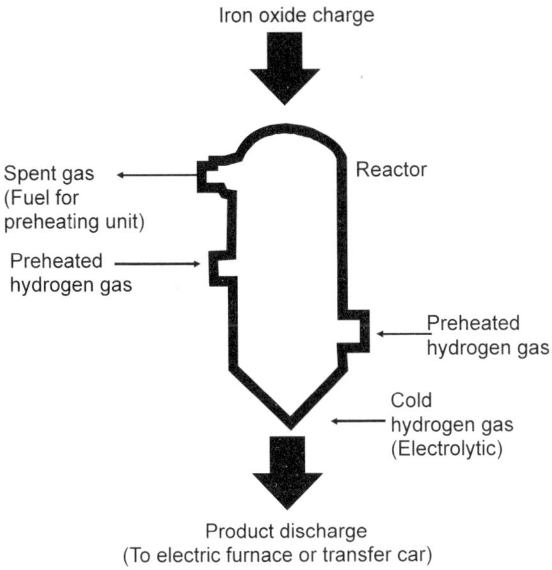

Fig. 15.5 Future iron making

these SR processes, not because they pose a threat to blast furnace but because they are being seen as forerunners of potential processes of the future. These can consume iron oxide feed, which cannot easily be taken up by either blast furnace or sponge iron making processes. Examples of these iron oxide feed are iron ore slimes, other iron oxide wastes from steel plants, lower grade iron ore fines accumulated at mine heads, Red Mud from aluminium extraction process, etc.

At present two Corex units are operating in India under the Jindal Vijayanagar Steel Limited (JVSL–now renamed JSW). The hot metal production is in excess of 1.5 mtpy. A 0.3 mtpy capacity plant based on Romelt Process was being considered by National Mineral Development Corporation (NMDC). Another Romelt unit of 0.243 mtpy capacity was being planned by National Aluminium Company (NALCO) in Orissa based on their waste product "red mud" which is basically iron oxide. Both the proposals have now been kept on hold.

Other smelting reduction processes have not been seriously considered in India till now. We, however, need to watch the developments related to CONTIFUR (acronym for continuous induction furnace) Process being developed by M/s Electrotherm (India) Limited in the western Indian state of Gujrat.

15.11 MODERN TRENDS IN IRON MAKING

In-spite of problems related to availability of hard coke, blast furnace iron making is still showing a reasonable growth. In-spite of the fact that coke cost may further increase, we may still see installation of more mini blast furnaces and thus further boost quantity of blast furnace iron. This is because large blast furnaces had high capital invested therein and need to operate at full capacity. Coke ovens catering to these blast furnaces have to continue to operate even at higher cost. In the process, coke ovens would be producing a fraction of coke, which cannot be used in the large blast furnaces, and these, mostly smaller size coke, will remain available for mini blast furnaces to use. Further, doubts about the availability of coke for the mini blast furnaces and sudden spurts in coke prices can be mitigated by partially using sponge iron in their charge, thereby reducing the quantity of coke usage.

As mentioned earlier, iron making in India is still dominated by blast furnaces, accounting for nearly 55% of the total iron production.

Availability of sponge iron has led to a situation in India where sponge iron making and blast furnaces are partners in industrial growth. As mentioned earlier in chapter 12, Fig. 12.2 schematically presents a typical concept, which newer mini steel plants of

India may adopt. In this concept, blast furnace iron making and sponge iron making have been judiciously married for economic viability.

Even though smelting reduction processes are considered forerunners of processes of the future, still their adoption in the near future remains uncertain and they are unlikely to occupy a dominant position.

15.12 A PEEP INTO THE IRON AND STEEL MAKING PRACTICE OF TWENTY SECOND CENTURY

Making iron and steel causes incremental damage to environment, which are cumulating over the years. Green house effect of accumulated carbon dioxide in the atmosphere is already affecting global climatic conditions. Strict restrictions are coming into force for emission control in automobiles and in production plants.

We expect that within this century the problem is likely to grow to the question of survival of human race. With this imperative in mind, the next century is likely to permit only those iron and steel making processes, which are safe for the environment. With this compulsion in mind, it should be possible to hazard a guess regarding the iron and steel making processes of the twenty-second century.

Environmental concerns would force minimisation of CO_2 emissions from iron and steel making. Possibly, only the CO_2 emitted in electric melting of scrap or equivalent may be permitted. About 40% of the steel making would be from recycled scrap. The balance metallic input has to come from DRI, which would be made from hydrogen reduction of iron oxide (Fig. 15.5). This hydrogen again may be mostly generated from the electrolysis of water. Most of the electric power needs of the entire steel making chain would come from non-CO_2 producing environment friendly route. Hydro electricity, sun power, wind power and geo-thermal power would all be contributors. Power from the waves of seashore may also be contributing.

Thus, even though there is major competition among the newer processes described earlier, and many of them may succeed commercially, their life is not going to be long after all. The story of iron and steel making had started with sponge iron and at the end it would still be the sponge iron, which would predominate, even though it would be made differently.

15.13 WHAT MAKES ROTARY KILN SPONGE IRON MAKING CLICK IN INDIA?

The reason for the above lies mainly in the type of coal available in India. Most of the Indian coals are such that the ash contributing mineral matter is well dispersed in the matrix. Thus most of the coals are high in ash, which cannot be beneficiated without large reduction in size. Grinding coal to very fine size renders it unsuitable for many applications. Earlier, very high ash coals were all diverted for use in thermal power plants for burning in the pulverised coal burners, but sponge iron making in rotary kiln is another reactor where high ash coals can be tolerated. Rotary kiln has large freeboard, which allows heavily dust-laden gases to pass without problem.

The sponge iron Pilot Plant of RDCIS SAIL could be operated for full one week on a coal with average ash content of 45%, which occasionally went as high as 53%. Although

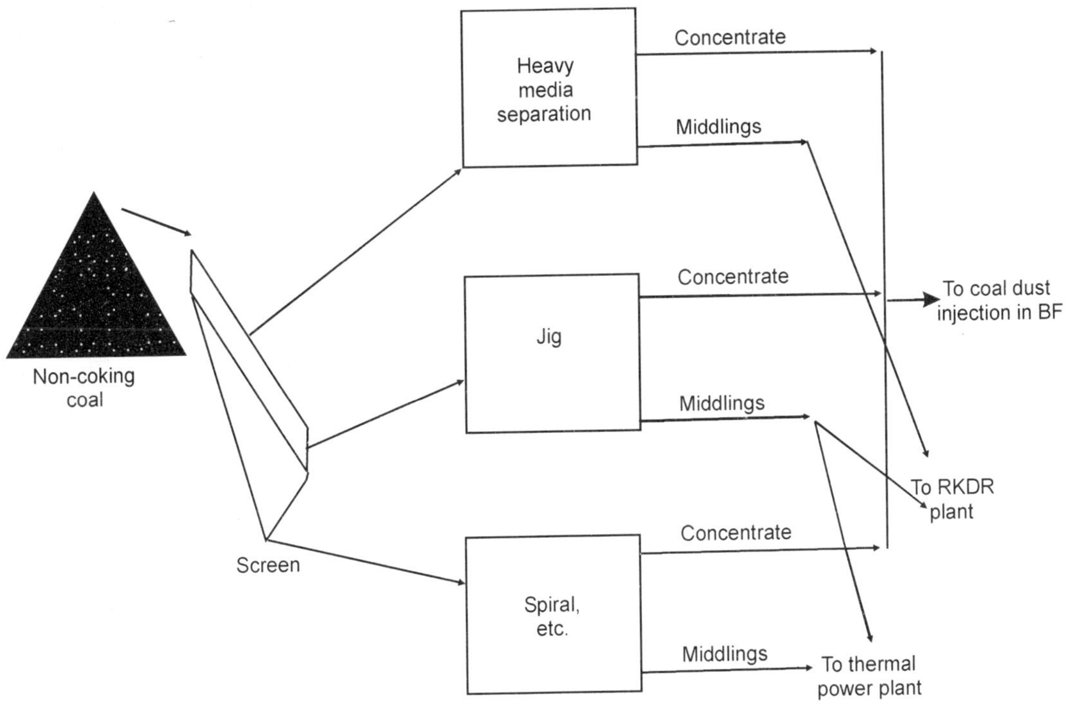

Fig. 15.6 Future scenario of outputs from coal washeries

productivity dropped sharply to less that 60% of the normal value, satisfactory metallisation level could be maintained.

Rotary kiln sponge iron making process appears to be tailor-made for Indian coals – and is poised to occupy centre stage in iron production in India.

In the coming years, technology policies of governments are going to be increasingly dominated by concern for the environment. Emphasis would be more on environment friendly technologies and here also rotary kiln sponge iron making has a definite edge over other coal based iron making processes. If the optimised concept of preheating of charge and of process air by waste gas is adopted, the edge could widen to an unsurpassable gap.

15.14 ANOTHER OPPORTUNITY FOR ROTARY KILN SPONGE IRON MAKING?

Another opportunity is waiting in the wings for rotary kiln sponge iron making. The blast furnaces in the integrated steel plant of India are presently using considerable quantities of imported coals taking advantage of low international prices. But attempts have been focussed in reducing this dependence on imported coal through the following measures:

1. Reducing consumption of metallurgical coke in blast furnaces
2. Using coke breeze along with sinter in blast furnace
3. Replacing metallurgical coke through coal dust injection through tuyeres

(using beneficiated indigenous non-coking coal)

4. Using indigenous blendable coal in the coke blend.

The option 3 listed above is a very important option, which is likely to be exercised, in stages, by most of the blast furnaces in India.

This would necessitate setting up of large capacity (or large number of) coal washeries for making available the lower ash coal needed for coal dust injection. The middlings generated from such coal washing plants may be of such quantity that it may not be possible to consume all of it in the traditional consuming point – the pulverised coal burners of thermal power plants. Rotary kiln sponge iron plants would have the opportunity to use this high ash coal fraction. If properly designed for using high ash coal, there is no metallurgical reactor which is better suited for handling high ash coal than the rotary kiln. The large free board in the sponge iron rotary kiln makes it amenable to handle abuse with large quantities of flying ash. The gas turbulence existing inside the sponge iron rotary kiln ensures that heat transfer is not hindered by increase in the dust load of free board gas.

In conclusion, we would like to present pictorially a scenario of coal utilisation by integrated steel plants, which is presented in Fig. 15.6. This Figure points towards the availability of massive quantities of coal middlings in the coming days. This is an opportunity, which the sponge iron industry can ill afford to miss.

Bibliography

- Abraham, M.C. and A. Ghosh, Kinetics of reduction of iron oxide by carbon, Iron making and Steel making, 7(1), 14-23 (1979).
- Agariya, Dalganjan and K.K. Prasad, Iron making technique of the Agariyas: In the words of an Agariya (in Hindi), Ispat Vikas (First issue of 1995) (Published by RDCIS, SAIL), 4-6, Ranchi.
- Agrawal, B.B. and K.K. Prasad, Evolution of sponge iron technology - Invention Intelligence, 23(8), 211-216 (August 1988), New Delhi.
- Agrawal, B.B., K.K. Prasad and G.I.S. Chauhan, Utilisation of iron ore fines to produce ore-coal composite pellets for improved sponge iron making in rotary kiln, Proc. Symp. Potential of sponge iron in Indian steel industry, Ranchi, 2005.
- Agrawal, B.B., K.K. Prasad, G.I.S. Chauhan, P.K. Chaudhuri & S.M. Aeron, Evaluation of earlier iron making processes and their relevance in Indian context, presented in the National Seminar on Indian Archaeometallurgy, Oct.1991, The Banaras Hindu University, Varanasi.
- AISE Steel Foundation (Pub.), The Making Shaping & Treating of Steel, 11th edition (1999).
- Ang, H.M., M.O. Tade & M.W. Sze, Residence time distribution for a cold model rotary kiln, AUSIMM Proceedings 303, (1), Nov. 11-16, 1998 (Australia).
- Balasubramaniam. D., Radio isotopic determination of charge movement in a Rotary Kiln, M. Tech. Thesis, IIT Kanpur, India, (Jan. 1981).
- Balasubramaniam. R., Story of the Delhi Iron Pillar, Foundation Books, New Delhi, 2005.
- Bandopadhyay, A., A. Ganguly, K.K. Prasad, S.B. Sarkar and H.S. Ray, Determination of Kinetic Parameters for the Re-oxidation of Direct Reduced Iron under rising temperature conditions, Thermochimica Acta, 228, 271-281 (1993).
- Bandopadhyay, A., A.K. Ray, K.K. Prasad, P.K. Bandyopadhyay and B.R. Chowdhury, Modelling of Heat Transfer in the Preheating Zone of a Rotary Kiln for Sponge Ironmaking, Indian Chemical Engineer, 31(4), 8-11 (Oct-Dec. 1989).
- Bandopadhyay, A., A.K. Ray, M.P. Srivastava, K.K. Prasad, S.V.B. Subba

Rao, P.K. Bandyopadhyay, R. Haque and B.R. Choudhary, Selection of Coals for Rotary Kiln Sponge Ironmaking, Trans. Indian Inst. Metals, 40(3), 209-218 (June 1987).

- Bandopadhyay, A., Amit Ganguly, K.K. Prasad, S.B. Sarkar and H.S. Ray, On the Determination of Ignition Temperature of Sponge Iron, Thermochimica Acta, 149, 27-31 (1989).
- Bandopadhyay, A., Amit Ganguly, K.K. Prasad, S.B. Sarkar and H.S. Ray, Low and High Temperature Re-oxidation of Direct Reduced Iron: A Relative Assessment, Reactivity of Solids, 8, 77-89 (1990).
- Bandopadhyay, A., Amit Ganguly, K.K. Prasad, S.B. Sarkar and H.S. Ray, Low and High Temperature Re-oxidation of Direct Reduced Iron: A Relative Assessment, Reactivity of Solids, 8, 77-89 (1990).
- Bandopadhyay, A., Amit Ganguly, K.K. Prasad, S.B. Sarkar and H.S. Ray, Thermo-gravimetric Studies on the Re-oxidation of Direct Reduced Iron, ISIJ International, 29(9), 753-760 (Sept.1989).
- Bandopadhyay, A., Amit Ganguly, K.K. Prasad, S.B. Sarkar and H.S. Ray, On the Determination of Ignition Temperature of Sponge Iron, Thermochimica Acta, 149, 27-31 (1989).
- Bandopadhyay, A., Amit Ganguly, K.K. Prasad, S.B. Sarkar and H.S. Ray, Protective techniques to combat sponge iron re-oxidation, Trans. Indian Inst. Metals, 44(4), 299-304 (August 1991).
- Bandopadhyay, A., M.P. Srivastava, A.K. Ray and K.K. Prasad, Effects of operating parameters on movement of charge in a rotary kiln for sponge iron making, Steel India, 11 (1), 9-17 (April 1988), Ranchi.
- Bandopadhyay, A., M.P. Srivastava, A.K. Ray & K.K. Prasad, Mathematical modelling of charge movement in rotary kiln for sponge ironmaking, Trans. Indian Inst. Metals, 39(3), 181-186 (June 1986), Kolkata.
- Bandopadhyay, Amitava, Amit Ganguly, K.K. Prasad, A.K. Singhi and S.K. Gupta, DRI on Fire: A Case Study, Steel India, 10(1), 58-63 (April 1987).
- Bandopadhyay, Amitava, Amit Ganguly, K.K. Prasad, P.K. Bandyopadhyay, S.B. Sarkar and H.S. Ray, Determination of Ignition Temperature of Sponge Iron Samples by Thermal Analysis, Proc. Sixth National Symp. Thermal Analysis, Delhi (Dec.1987), V16-V21.
- Bandopadhyay, Amitava, Amit Ganguly, K.K. Prasad, S.B. Sarkar and H.S. Ray, Newer Approaches to the Study of Re-oxidation of Sponge Iron - Trans. Indian Inst. Metals, 42(1), 21-26 (Feb. 1989).
- Bandyopadhyay, D., N. Chakraborti and A. Ghosh, A Study on the Kinetics of Iron Oxide Reduction by Solid Carbon, Steel Res., 64 (7), 340-345 (July 1993).
- Barr, P.V., J.K. Brimacombe and A.P. Watkinson, Heat transfer model for rotary kiln: Part I, Metall. Trans. B, 20B, 391-402 (1989).
- Barr, P.V., J.K. Brimacombe and A.P. Watkinson, Heat transfer model for rotary kiln: Part II, Metall. Trans. B, 20B, 403-419 (1989).

- Batelle Institute e.v., Frankfurt, Report on Reduction processes outside the blast furnace and their effect on future iron and steel production in the world, 1975.
- Bayard, R.A., Chemical and Metallurgical Engineering, 52, p.100 (1945).
- Biswas, D.K., Some design aspects of pneumatic coal injection for coal based direct reduction process in rotary kiln, Powder Handling and Processing, 5(2), 145-152 (June 1993).
- Cartaxo, S.J.M. and S.C.S. Rocha, Object oriented simulation of pneumatic conveying: Application to a turbulent flow, Braz. J. Chem. Engg. 16(4), Dec. 1999.
- Chatterjee, A., A.V. Sathe, M.P. Srivastava & P.K. Mukhopadhyay, Met. Trans. B. 14B, 375-381 (September 1983).
- Chatterjee, A., Beyond the blast furnace, CRC Press, USA, 1994.
- Chatterjee, A. and B.D. Pandey, Iron making and Steel making, Mechanism of ring formation in rotary kilns used for sponge iron manufacture, 8(6), 250-263 (1981).
- Chatterjee, A., Direct reduction of iron ores, Prentice Hall, New Delhi, 2008.
- Chatterjee, A. and A. Ghosh, Iron making and steel making. Theory and Practice, Prentice Hall, New Delhi, 2008.
- Chen, Y., H. Zhang and Y. Xie, Study on reduction mechanism of chromium ore pellets mixed with carbonaceous breeze reductants, Sel. Pap. Eng. Chem. Metall. (China), 41-49 (1993).
- Choudhary, S.K. and A. Ghosh, Study of coal reactivity with reference to direct reduction processes, Trans. Indian Inst. Metals, 42 (6), 541-552 (Dec. 1989).
- Crawford, G.P., Iron & Steel Making in Soderberg electric arc furnace, Iron making & Stee lmaking, 16 (5), 314-319 (October 1989), London.
- Dey, Shobhana, Flotation responses of Gondwana coal in release analysis, Mineral Processing and Extractive Metallurgy, 114(4), 248-254, Dec. 2005.
- Direct from Midrex, Quarterly publication from Midrex Corporation, U.S.A., various issues since 1993, Charlotte.
- Feinman, Jerome and Donald R. Mac Rae (Ed.), Direct Reduced Iron: Technology and economics of production and use, Iron & Steel Society, Warrandale, P.A., 1999.
- Ganguly, A. and A. Bandopadhyay, Self-ignition tendency – A hazard for sponge iron producer and user, Tool & Alloy Steels, 22(1), 19-25 (1988).
- Ginstling, A.M, D.M. Zil'berman & N.V. Gvozdev, Khim. Mashinostroenic, 8, p.8 (1939); as quoted by Friedman, S.J. & Marshall, W.R., Jr., Chemical Engineering Progress, 45 (8), pp. 482 493 (1945).
- Gupta, A.K., Ed., Technology of sponge iron, Metallurgical Research Institute, Roorkee, India, 1979.
- Hobhouse, Steve, Iron and Steel Making at New Zealand Steel, JPC Bulletin, Kolkata, 5(1), 72-75 (January 2005).

- Jiang, Tao, Guanzhou Qiu, Jingkang Xu, Deqing Zhu and Ram Singh, Direct reduction of composite binder pellets and use of DRI, Electrotherm (India) Ltd., 72-Palodia, Ahmedabad, India.
- Kong, L., T. Martinovic and W.K. Lu, Reducibility of DR pellets, 58th Iron making Conference Proceedings 1999, 383-390 (Mar. 1999).
- Kubaschewski, O. and C.B. Alcock, Metallurgical Thermochemistry, 5th Edition, Pergamon Press, London, 1979.
- Kumar, S.K., B.B. Agrawal, T.K. Datta, K.K. Prasad, M.S.R. Iyengar & H.S. Ashwath, Impact of Sponge Iron on the Growth of Steel Industry, Proc. Symp. Use of Sponge Iron in Electric and Arc Melting Units, Hyderabad, 295-312 (July 1984).
- Levin, Ernest M., Carl R. Robbins & Howard F. McMurdie, Phase Diagrams for Ceramists, 3rd edition, American Ceramic Society, Columbus, Ohio, 1974.
- Lowry, H.H. (Ed.), Chemistry of Coal Utilisation, Supplementary volume, 521-522 (1976).
- Mazumder, A.K., T. Sharma & G.V. Reddy, Reduction kinetics of iron oxide pellet by non-coking coal, Trans. Indian Inst. Metals, 42 (3), 299-305 (June 1989).
- Mott, R.A. and R.V. Wheeler, The quality of coke, Chapman and Hall, London 1939, p. 464.
- Mukhopadhyay, P.K., A.V. Sathe and Amit Chatterjee, A Mathematical Model of Direct Reduction of Iron Oxides in Rotary Kilns using the TDR Process, Trans. Indian Inst. Metals, 37(6), 721-729 (December 1984).
- Prakash, S. and H.S. Ray, Reduction of iron ore under rising temperature and fluctuating temperature conditions, Thermochimica Acta, 111, 143-166(1987).
- Prasad, K.K., Iron making by indigenous communities in India, Part I: The scene and the Orissa and Bishunpur Processes, Steel Times, 227 (12), p. 461 (December 1999), London.
- Prasad, K.K., Iron making practice of indigenous communities in India, Part II: Embers survive, Steel Times, 228 (1), p. 35 (January 2000), London.
- Prasad, K.K., Traditional ironmaking by indigenous communities of India, Ironmaking & Steelmaking, 27 (6), 408-410 (December 2000), London.
- Prasad, K.K., Residence time of solids in a sponge iron rotary kiln: Modelling and measurement techniques, Proceedings of National Seminar on Experimental Approaches in pyrometallurgical research, Regional Research Laboratory, Bhubaneswar, India, April 20-21, 2000, pp. 85-96, Allied Publishers, New Delhi.
- Prasad, K.K., Steel Demand Forecasts: India 1999-2000, Steel Times International, 23 (4), pp. 14-17 (July '99), London.
- Prasad, K.K., Alternate Iron Making - India is the World Leader, Steel and Metallurgy, 6(3), 41-44 (January 2004).
- Prasad, K.K., Sponge Iron in Indian Iron and Steel Industry, JPC Bulletin, Kolkata, 4(5), 11-18 (May 2004).

- Prasad, K.K., Key issues in the Production and Utilisation of Sponge Iron, Proc Symp. Sponge Iron Industry – Problems and Prospects, Bhubaneshwar, 7-20 (June 25-26, 2004).
- Prasad, K.K., Implications of charging sponge iron in blast furnace, Steel and Metallurgy, 6(10), 12-17 (August 2004).
- Prasad, K.K., What the future holds for iron making – Newer processes, Steel Times International, 26 (8), 30-34 (September 2004).
- Prasad, K.K., Auto-ignition of Sponge Iron - A Review, Steel and Metallurgy, 7(3), 16-23 (January 2005).
- Prasad, K.K., Meeting iron needs of India: Next four decades, Proc. Symp. Iron making: Technology and Raw Materials, Kolkata, 26-36 (July 1-2, 2005).
- Prasad, K.K., Sponge Iron: Key to India's development in coming decades, JPC Bulletin, Kolkata, 5(9), 6-13 (September 2005).
- Prasad, K.K., Sponge Iron as a Raw Material for Iron and Steel Making, JPC Bulletin, Kolkata, 5(11), 19-25 (November 2005).
- Prasad, K.K., Mysteries of the Delhi Iron Pillar, Part 1: History and Properties, Steel Times International, 30 (5), 52 (July/ August 2006).
- Prasad, K.K., Mysteries of the Delhi Iron Pillar, Part 2: Making the Pillar, Steel Times International, 30 (6), 70 (September 2006).
- Prasad, K.K., Iron and steel making: How we reached there, Everyman's Science, 41(4), 249-254 (October-November 2006).
- Prasad, K.K. (Ed.), Steel Plant Technologies: An overview, RDCIS SAIL, Ranchi, 2004.
- Prasad, K.K., A. Bandopadhyay & P.K. Bandopadhyay, Sponge Ironmaking in Rotary Kiln: Strengths and Weaknesses, Invention Intelligence, 23(10), 313-316 (October 1988).
- Prasad, K.K., B.B. Agrawal, & H.D. Pandey, Some environment aspects of coal based Sponge Iron making in India, Proc. Symp. Environmental Management in Metallurgical Industries, Institute of Technology, Banaras Hindu University, Varanasi, India, December 14-16, 2000, pp. 275-281.
- Prasad, K.K. and S.R. Mediratta, Forecasting Demand for Steel, Steel Technical Report, November '97, pp.16-23, Lucknow.
- Prasad, K.K. and A.K. Ray, Sponge Iron Making in Rotary Kiln, R&D Centre for Iron & Steel, SAIL, Ranchi, April 2002.
- Prasad, K.K. and H.S. Ray, The making of (and attempts at breaking) the iron pillar of Delhi, Steel World, 51-56, June 2001.
- Prasad, K.K. and H.S. Ray, A pillar of the community from the first to the third millennium, Materials World, 8(7), 11-13, July 2000.
- Prasad, K.K., H.S. Ray & K.P. Abraham, Chemical and Metallurgical Thermodynamics, New Age International (P) Ltd., Publishers, New Delhi, 2007.
- Prasad, K.K., S.P.S. Sabharwal, M.P. Srivastava, B.B. Agrawal, G.I.S. Chauhan, Mahesh Sharma, P.K. Chaudhuri, S.M. Aeron and C.R. Srinivasan. An attempt

on revival of rural iron making, Metal News of Indian Inst. Metals, 12(1), 1-6 (Feb.1990), Kolkata.

- Ramakrishnan, V. and P.S.T. Sai, Mathematical modelling of pneumatic char injection in a direct reduction rotary kiln, Met. Trans. B, 30B, (5), 969-977B (Oct. 1999).
- Ray, A.K., K.K. Prasad and P.K. Sen, Residence-time in rotary kiln for sponge iron making: Prediction from kinetic data, Steel India 17(1), 16-20 (April 1994), Ranchi.
- Ray, Asim Kumar, Ph.D. Thesis, IIT Kharagpur (1988).
- Ray; A.K., K.K. Prasad, H.S. Ray and P.K. Sen, Residence time in a rotary kiln for sponge iron making: Prediction from kinetic data, Steel India 17 (1), 16-20 (April 1994).
- Ray, H.S., Kinetics of Metallurgical Reactions, Oxford & IBH Publishing Co., New Delhi, 1993.
- Ray, H.S., Non-isothermal kinetics – A generalised approach and simulation of some industrial reactors, Proc. Symp. Thermodynamics and Kinetics of Metallurgical Processes (ICMS-81, Bangalore), Indian Institute of Metals, Calcutta (1985).
- RDCIS SAIL, Trends in beneficiation and Agglomeration of iron ore, Ranchi, 2001.
- Reuter G. and H. Serbent, Raw material selection criterion for the SL/RN Process, Proc. Symp. The utilisation of pre-reduced material in iron & steel making, Bucharest, May 1976.
- Saeman, W.C., Passage time of solids through rotary kilns; Factors affecting time of passage, Chemical Engineering Progress, 47 (10), 508–514 (1951).
- Santra, A.K., A. Rajasekharan, V.M. Daptardar & V.L.N. Murthy, Statistical Modelling of Rotary Kiln Process, Presented in the 38th ATM of IIM, Bangalore, (November 1984).
- Schnabel, Wolfram, On the heat transfer during direct reduction in a rotary kiln, Ph.D. Thesis, Rheinisch-Westfaelischen Technischen Hochschule, Aachen, 1977.
- Serbent, H. & R. Dayal, Basic principles, process characteristics, and industrial application of rotary kiln iron ore reduction, Trans. Indian Inst. Metals, 30(4), 246-255 (August 1977).
- Sharma, T., Reduction of Iron Ore Pellets by Char Fines, Trans. Indian Inst. Metals, 43 (6), 380-382 (Dec. 1990).
- Singh, R.K., A.K. Ray & K.K. Prasad, CFD Analysis of combustion air flow inside a sponge iron rotary kiln, Steel Tech Journal, Kolkata (in press).
- Sinhamahapatra, A.K., K.N. Gupta, S.D. Singh & M. Ram, Kinetics and Mechanism of Formation of Direct Reduced Iron Rod from Ore and Coal Fines, Proceedings of 3rd International Symposium on Beneficiation and Agglomeration, Bhubaneswar, India, 506-512 (16-18 Jan. 1991).
- Srinivasan, N.S. and A.K. Lahiri, Rotary Kiln Process: A simplified analysis, Trans. Indian Inst. Metals, 30(5), 309-312 (October 1977).

- Srinivasan, Sharada and S. Ranganathan, India's legendary Wootz steel, IISc, Bangalore, 2004.
- Stephenson, R.L. (Ed.), Direct Reduced Iron: Technology and economics of production and use, Iron & Steel Society of AIME, Warrandale, P.A., 1980.
- Sullivan, J.D., C.G. Mair and O.C. Ralston, U.S. Bureau of Mines technical papers. 384 (1927).
- United States Steel (Pub.), The Making Shaping and Treating of Steel, 10th edition, The Association of Iron & Steel Engineers (1985).
- Vahl, L. and W.G. Knigma, Transport of solids through horizontal rotary cylinders, Chemical Engineering Science, 1, 253 (1952).
- Venkateswaran, V. and J.K. Brimacombe, Met. Trans., 8B, 387 (1977).
- Vishwakarma, Sonadhar Poyam and K.K. Prasad, First iron smelting: A Lohra folktale from Bastar, Invention Intelligence, 30(10), 9-10 (January 1955), New Delhi.
- Wingfield, S.L., A. Prothero and J.B. Auld, Jl. Inst. Fuels, (1974), 64.
- Zablotny, W.W., The movement of charge in a rotary kiln, International Chemical Engineering, 5(2) 360-366, April 1965.

Index

D

E

S

T